GEEK

GIRLS

UNITE

ALSO BY LESLIE SIMON

Everybody Hurts: An Essential Guide to Emo Culture (with Trevor Kelley)

Wish You Were Here: An Essential Guide to Your Favorite Music Scenes—from Punk to Indie and Everything in Between

GEEK GIRLS UNITE

How Fangirls, Bookworms, Indie Chicks,
and Other Misfits Are Taking Over the World

Leslie Simon

Illustrations by Nan Lawson

*it***books**

AN IMPRINT OF HARPERCOLLINS*PUBLISHERS*

*it***books**

HarperCollins books may be purchased for educational, business, or sales promotional use. For information please write: Special Markets Department, HarperCollins Publishers, 10 East 53rd Street, New York, NY 10022.

FIRST EDITION

Designed by Justin Dodd

Library of Congress Cataloging-in-Publication Data has been applied for.

ISBN 978-0-06-200273-0

11 12 13 14 15   OV/RRD   10 9 8 7 6 5 4 3 2 1

*To all the bands that have appeared on the soundtrack of my life,
and all the songs that have gotten me through the awesome times,
the awful times, and all the stuff in between*

CONTENTS

GEEK GIRLS UNITE

INTRODUCTION

I won the Invention Convention in second grade for creating Stegosaurus Cereal, a breakfast snack that was packed with nutrients *and* educational tidbits about our now-extinct terrestrial vertebrate friends. In seventh grade, I wrote the lyrics to the Eagles' "Hotel California" and Don McLean's "American Pie" on flash cards so I could memorize them and then sing along effortlessly when they were played at bar and bat mitzvahs. When I was twelve, I used to hold a tape recorder up to my television speakers to record episodes of the original *Beverly Hills, 90210* so I could listen to them while I sat by the pool. I keep a stash of Hello Kitty Band-Aids in my purse at all times. When I'm waiting for the subway, I often pass the time by trying to brainstorm new portmanteaus.* I wouldn't dare upgrade my cell

---

* Words that are the combination of two words blended together like "spork" ("spoon" + "fork"), "skort" ("skirt" + "shorts"), "Speidi" ("Spencer" + "Heidi"),

phone without consulting Bonnie Cha's reviews on CNet.com. I have a long-standing date with *Gilmore Girls* every weekday at 5:00 p.m. on ABC Family. Most of my Facebook status updates are Liz Lemon quotes from *30 Rock*. One of my goals in life is to collect Bright Eyes's entire discography. In other words . . .

My name is Leslie—and I'm a geek.

GEEK IS THE NEW COOL

Once upon a time, to be labeled a "geek" was a fate worse than death. It meant you were an outcast. A loser. Destined for a solitary existence where a twelve-sided die would provide you with the only action you'd ever know. However, over the past decade or so, four-eyed social pariahs have been waging a quiet—yet powerful—revolt. We've developed our own fashion sense. (Geek chic, anyone?) We've penned pieces of literary brilliance. (Rest in peace, David Foster Wallace.) We've even reinvented the art of modern cinema. (Kudos, Coen brothers!)

"Geek" is no longer a four-letter word. Instead, it's a badge of honor for anyone who has ever played World of Warcraft on a Friday night, appointed themselves commissioner of their fantasy football team, or memorized every single line of Jason Schwartzman's dialogue in *Rushmore*, using flash cards or not. It's a geek's world; everyone else is just living in it.

"But, Leslie," you cry, putting your hand on your hip and stomping your foot. "What's the difference between a geek and a nerd? Are geeks cooler than dorks? This is too confusing. I feel light-headed and my left arm is starting to go numb."

Take a deep breath. It's going to be all right, assuming the numbness isn't a warning sign of a stroke, in which case you should throw down this book and call 911 STAT. Let's outline

---

and a bunch of other words that don't start with the letter "s."

some basic terms that will help you understand what it means to be a geek (at least in the context of this book), as opposed to your typical nerd, dork, or weirdo.

geek (\ˈgēk\, n.) A person who is wildly passionate about an activity, interest, or scientific field and strives to be an expert in said avocation. Person does not necessarily sacrifice social status to participate in area of expertise; instead, person will often seek out like-minded peers—in both the real and the virtual worlds—in order to connect, bond, and celebrate mutual love for this area.

nerd (\ˈnərd\, n.) A person who excels academically and who thrives on such educationally induced pastimes as memorizing UNIX manuals and correcting your grammar. Such persons may not possess the most advanced social skills, but they are armed with a huge heart and an even bigger brain.

dork (\ˈdork\, n.) A person who is delightfully oblivious to present-day trends, slang, and references.

dweeb (\ˈdwēb\, n.) A person who is oftentimes mistaken for being highly intelligent when, in fact, the person is usually technologically and academically inept. The person may feign aptitude by doing such things as wearing a NO, I WILL NOT FIX YOUR COMPUTER T-shirt or talking in a robot voice.

weirdo (\'wir-(,)dō\, n.) A person who exhibits particularly strange, nonconformist, and eccentric behavior and mannerisms. *See* Bai Ling or Tila Tequila.

HE-GEEK VS. SHE-GEEK

Self-proclaimed geeks like actors Seth Rogen and Michael Cera have become inadvertent sex symbols. Tech-savvy savants like MySpace's Tom Anderson and Facebook's Mark Zuckerberg are probably laughing all the way to the bank—and flipping off all the high school jocks who made their lives miserable during their teenage years. I couldn't be happier that male geeks have already started to earn well-deserved respect in the eyes of pop culture. But what about their female counterparts?

Embracing the idea of "girl power" isn't something that just ceased to exist after the Spice Girls broke up . . . then got back together . . . and then broke up *again*. After all, fangirls are just like fanboys—we put on our Imperial Stormtrooper Lycra pants one leg at a time. But, for some reason, she-geeks have yet to be truly encouraged and accepted by mainstream society—and that's probably because we also remain largely misrepresented and misunderstood.

Contrary to popular belief, if you see a girl buying a copy of *BioShock 2*, that doesn't mean she's getting it for her little brother or her boyfriend. When you hear a girl say she enjoys "vampire lit," that doesn't mean she goes to sleep every night underneath her *Twilight* duvet cover. Oh, and eleven-year-old boys aren't the only ones who're stoked about the new *ThunderCats* cartoon reboot. Sure, gals like Google VP Marissa Mayer and explorista Samantha Brown have started to break new geek ground, but we've got a long way to go, ladies.

It's time for us to reclaim the connotations of being a "geek" and hold tight to the term as a source of pride and distinction.

In other words, embrace your quirkiness! Celebrate your idiosyncrasies! There is power in your geekiness! Trust.

MEET THE GEEK GIRL GUILD (AKA ΓΓΓ)

Because this is a topic near and dear to my heart, I wanted to do something I've never done before: involve readers in the writing process. More specifically, I wanted to engage with the kindred spirits I'd be applauding in the pages of this book. I never joined a sorority in college, so it seemed like a great idea to start my own as a way to connect with my long-lost, like-minded sisters. And so, the Geek Girl Guild was born.

After I posted an initial query on my website and forwarded the idea to a few friends to distribute, I was completely overwhelmed by the response. Women of all ages, backgrounds, and areas of geek expertise wanted to joined the sisterhood, making the first ΓΓΓ pledge class over one hundred strong! And these girls have as much spirit as any chapter of Pi Beta Phi. As part of their ΓΓΓ membership, they let me pick their geeky brains about a variety of topics so I could better understand what makes a she-geek tick. You'll find their incredible input peppered through the pages of *Geek Girls Unite*, which I couldn't have completed without their advice and insight.

ΓΓΓ

My goal for the Geek Girl Guild was to learn more about our individual passions, unite us as a geektastic girl community, and ultimately celebrate our fantastic uniqueness—and I think the Tri Gams achieved all this and more!

I hope you're ready, because you're about to run a geek girl marathon. Feel free to take a short break if, at any point in time, you start to cramp up. While you take a second to stretch and fill up your water bottle, let's preview some of the geek girls we're going to pass along the way.

The Fangirl Geek: When it comes to World of Warcraft, Neil Gaiman, and naming all the ancillary Hello Kitty characters, she's no n00b. It's no surprise she's rolled her twelve-sided die right into the center of our hearts.

The Literary Geek: This is the gal who always has her nose in a book, uses her library card on a regular basis, and actually thinks Chuck Klosterman is halfway attractive.

The Film Geek: She might be a little pale from spending so much time indoors and know *far* too much about French New Wave cinema, but she's got a black belt in Hollywood escapism and the Netflix queue to back it up.

The Music Geek: She prefers to sleep in her size XXL Smashing Pumpkins ZERO T-shirt, takes all Pitchfork album reviews with a grain of salt, and still keeps a candle lit in hopes of a possible Smiths reunion.

The Funny-Girl Geek: She's scrappy and has sass for days. In fact, she packs a one-two comedic punch that'll leave you seeing stars. Our snarky sisters love to laugh—whether it's *with* you or *at* you.

The Domestic Goddess Geek: She's the one with a not-so-slight girl-crush on Martha Stewart who insists on postponing plans that threaten to coincide with *Top Chef*. She can also tailor a T-shirt by using only a needle and dental floss—plus, don't get me started on her eye for interior design and mad culinary skillz.

Finally, let's not forget to wave hello to our **Tech, Fashionista, Political, Retro,** and **Athletic Geek Girls** as we cross the finish line.

GEEK GIRLS, UNITED WE STAND

Let's be totally honest: we are all guilty of excluding people who aren't like us. Whether or not we mean to, we've all done it at some point or another—whether it is out of our own insecurities and fears, or just being oblivious to the fact that someone else is there. So I apologize in advance if I've left out any area of geekdom that might be *your* particular area of expertise, because there are plenty of awesome obsessions that I wasn't able to include here. Here's the most important thing, though: just because our passions aren't the same,* that doesn't mean we aren't united in our geeky affection for whatever it is that makes us happy—even if it feels like society sometimes pits us against one another. Whether you're obsessed with Renaissance fairs or roller derby, scrapbooking or ska, being a geek should unite—not divide—us. If one good thing comes out of this book, it will be that you get to know your geeky sisters (and cousins) so you can recognize these fabulous ladies when you see them, start a conversation, and realize that our differences are actually what bring us closer together. And without further ado, let the geeky life-building begin!

---

* Hey, I don't expect everyone to love the Wombats and Maxïmo Park the same way I do!

One
FANGIRL GEEK

Are you able to tell the difference between Badtz-Maru and Chi Chai Monchan?* Back slowly away from the Sanrio store, rifle through your bag for some sort of writing implement, and test your fangirl geek knowledge!

1. Director George Lucas made a very controversial change to a scene in *Star Wars Episode IV: A New Hope*, where Greedo comes to the Mos Eisley Cantina to collect the bounty on Han Solo's head. In the 1997 Special Edition version, Greedo fires his blaster at Solo first and misses, and then Solo retaliates

---

* Both are beloved Sanrio characters related to Hello Kitty. Badtz-Maru is a male penguin with spiky hair who can often be found sticking out his tongue or blowing a raspberry, while Chi Chai Monchan is a male monkey who likes to balance bananas on his head.

by shooting Greedo. However, in the 1977 original film, Solo shoots Greedo without the Rodian bounty hunter ever firing a shot. Fans upset with the change can often be found uttering the phrase:

    A. "Greedo shot first."

    B. "A bird in the Han is worth two in the Bossk."

    C. "Han shot first."

2. In *Dr. Who*, the longest-running sci-fi drama in history, the doctor travels through time on his spacecraft, which is called the *TARDIS*. What is TARDIS an acronym for?

    A. Taking Apart Rational Distance If Stimulated.

    B. You really shouldn't say the word "tardis." It's not very politically correct. They preferred to be called "special" or "challenged."

    C. Time and Relative Dimensions in Space.

3. Which of the following Neil Gaiman books has *not* been made into a movie?

    A. *Stardust*.

    B. *Coraline*.

    C. *The Day I Swapped My Dad for Two Goldfish*.

4. What is the Honor System in World of Warcraft?

    A. A code of practice based on trust and honesty.

    B. Don't you mean Honor Society? They're a pop-rock band on the Jonas Brothers' record label, Jonas Records. *Duh*.

    C. It allows WoW players to gain Honor Points, which are earned through Battlegrounds and World Outdoor PvP and are based on PvP kills and battles. Players can then spend said Honor Points on special rewards, like equipment and weapons.

5. In 1988, comic book artist Jamie Hewlett debuted the *Tank Girl* series, which centered around a foul-mouthed, drug-addled female outlaw with a multimillion-dollar bounty on her Mohawked head. Though she spent most of her time running around postapocalyptic Australia (and then postapocalyptic Britain) with her boyfriend Booga, a mutant kangaroo, she would go on to become an anarchist antiheroine. Hewlett was involved with the comic until 1996, after which he started the first "virtual band" made of comic book characters. That band is called:

    A. Wolfmother.

    B. The Monkeys.

    C. Gorillaz.*

ANSWER KEY

Mostly As: Your gumption is greatly admired, but you're going to have to do a lot more than watch *Coraline* in 3-D to be considered a true fangirl geek. Don't worry, though. By the end of this chapter, you'll be bursting with fangirl wisdom.

Mostly Bs: Sorry, Heidi Montag, but just because your body is 70 percent synthetic material doesn't mean you have anything in common with the kind of plastic used to make designer toys for Kidrobot and Tokidoki. In fact, don't you have to attend the opening of an envelope right about now? I won't be offended if you have to excuse yourself. In fact, I insist.

Mostly Cs: Major props, mademoiselle. You're a bona fide fangirl geek!

---

* Hewlett created Gorillaz with Blur frontman Damon Albarn, and the group's eponymous debut went on to sell seven million copies, earning them an entry in the *Guinness Book of World Records* as the Most Successful Virtual Band.

CHARACTER SKETCH

There comes a time in almost every girl's life when she outgrows her toys. Her collection of American Girl dolls and accessories disappears from her bedside table and begins a mass migration to the attic, along with her Bratz, Barbies, and all their teeny-tiny high heels with them. She asks for the Project Runway Make-up Artist Studio Box for her next birthday. She stops playing *Super Mario Bros.* on her Nintendo DS in favor of watching the Jonas Brothers in *Camp Rock*, and before you know it, she's one pack of Justin Bieber Silly Bandz away from becoming a dreaded tween.

The thing that separates the fangirl from the average girl is that she never really outgrows her toys. Instead, her girlhood obsession[*] becomes more serious and intense with age. For fangirls, collecting Hello Kitty cookware, rocking along to Jem & the Holograms, or reading about Archie Andrews is more than just a hobby; it's an unfettered source of happiness—and why would anyone want to put an end to their happiness just because they've reached an arbitrary age ceiling? That just seems cruel and inhumane.

The teenage years can be tough on the fangirl, though. Clothes and boys excite her classmates, while she's more stoked by a marathon screening of *Dexter*. Her peers can't wait for the weekend so they can go to school dances or drink Boone's Farm behind the bleachers, but her favorite day of the week is Wednesday because that's when all the new comics come out. For these reasons, fangirls often rank somewhere between the euphonium player in the marching band and the drama club lighting designer on the high school popularity totem pole. It's

---

[*] I realize that some girls don't discover their passions until later in life, but a good number *do* credit their favorite childhood pastimes with awaking their inner fangirl.

no wonder that most fangirls have a hard time relating to their contemporaries and often choose to participate in online communities rather than partake in real-world socialization. After all, by creating an avatar, fangirls are free to be whoever they *are*—or whoever they *want* to be—without the fear of mockery, misunderstanding, or the impending threat of being shoved into a locker.

> "Never be ashamed! There's some who'll hold it against you, but they're not worth bothering with."
> —author J. K. Rowling

SAY WHAT? THE FANGIRL LEXICON

It would be tough to bond with a fangirl geek without mastering the proper syntax. After all, you don't want to tweet something like ROTFL* when you really meant to type RTFM.† *That* would be embarrassing. In order to save yourself from a future FML‡ moment, I've compiled a glossary of essential fangirl terms so you can mingle seamlessly with the manga masses.

Anime (n.) Abbreviation for "Japanese animation." The drawing style is typified by panel layouts, exclamatory dialogue, and characters that have exaggerated features and lip-synch worse than Ashlee Simpson.

Avatar (n.) A computer-generated representation of a gamer.

Cosplay (n.) Abbreviation for "costume play," as when a fan dresses up like her favorite movie, comic book, and/or anime character for a convention, to go LARPing (see below), or, for those with less of a social life, to go grocery shopping.

Guild (n.) A video-gaming clan or posse.

IRL (adj.) Abbreviation for "In Real Life."

---

* Abbr. for "Rolling On The Floor Laughing."
† Abbr. for "Read The F***ing Manual."
‡ Abbr. for "F*** My Life."

LARP (n.) Abbreviation for "live-action role-play." To dress up as your favorite gaming, comic book, and/or anime character and act out said character's actions. LARPing is usually a group activity and involves various characters in a fictional setting, all interacting with one another and following a set of organized game rules.

Manga (n.) A Japanese comic book or graphic novel.

MMORPG (n.) Abbreviation for "massively multiplayer online role-playing game," which pertains to games like Final Fantasy and World of Warcraft.

Muggle (n.) Nickname for a *Harry Potter* fanatic.

NOOb (n.) Someone new to a gaming, comic book, or virtual community. Also known as a "newbie."

Pwn (pronounced pôn) (v.) To kick butt, dominate, or own an opponent.

RPG (n.) Abbreviation for "role-playing game."

Squee (n.) The sound a fangirl makes when she's really, *really* excited. A squee is somewhere between a squeal, a scream, and a cry and can reach upward of 130 decibels—especially if Taylor Lautner is involved.

Tolkienist (n.) Someone who studies the works of J. R. R. Tolkien.

Trekkie (n.) Nickname for a *Star Trek* fanatic.

Twi-hard or Twilighter (n.) Nickname for a *Twilight* fanatic.

Whovian (n.) Nickname for a *Dr. Who* fanatic.

GEEK MYTHOLOGY

Some cite **Margaret Cavendish,** Duchess of Newcastle, as writing the first book of fantasy, entitled *The Blazing World*, which was about a Utopian world discovered by way of the North Pole in 1666, but the majority of horror historians credit **Mary Shelley** with inspiring the literary genre of science fiction with her publication of *Frankenstein* in 1818. The novel, which centers on a scientist who creates the likeness of man (in the most monster-y sense), was a huge success and forged a path for future horror/sci-fi writers like Edgar Allan Poe and H. G. Wells.

Aliens, zombies, and werewolves have always been a fangirl fascination, and no one has unearthed the undead's secrets quite like author **Anne Rice**, who rose to prominence in the 1970s with her *Vampire Chronicles* series, including titles like *Interview with the Vampire* and *The Queen of the Damned*. Unlike previous literary and cinematic depictions of vampires, Rice modernized and redefined these misunderstood members of the undead by making them sensitive, highly intelligent, and very, *very* sexual. Nearly thirty years later, a new generation of vampires would take a bite out of pop culture, but this time they would choose romance over sex. (A bold concept!) Yes, **Stephenie Meyer's** *Twilight* saga ignited a worldwide obsession with self-restrained bloodsuckers—and, more important, reading. (An even *bolder* concept!) Other pieces of supernatural fiction followed—like **Suzanne Collins's** *Hunger Games* trilogy or **Sophie Jordan's** *Firelight* series—and it seems like you can't wake up without hearing about a new franchise of dragon-hunting, vampire-slaying, werewolf-fighting heroines.

While survivalist female characters have been relatively easy to find in fantasy and science fiction, it took a bit longer for strong women to emerge in the comic book world. In fact, comic book writers have long been guilty of depicting women as little more than pieces of overdrawn eye candy. When comics first gained popularity in the 1930s, women were depicted as (1) weepy, big-bosomed damsels in distress, (2) fun-loving, big-bosomed teenagers, or (3) icy, big-bosomed heroines. It wasn't until 1954 that the Comics Code was implemented by the Association of Comic Magazine Publishers, stating, "Females shall be drawn realistically without exaggeration of any physical qualities." This was great news with regard to

"I call myself a 'trekker,' because 'trekkie' means you love Kirk (who I despise) and it also means you wear Spock ears on a regular basis. Trekkers are in love with the philosophy of Gene Roddenberry's *Star Trek*. He created a world where there is no currency, no disease, limited greed, [and] limited war. The galaxies are not looking to gain material possessions, but rather they are exploring and attempting to learn more about the universe and the beings in that universe. What a lovely concept, eh?"

Lindsay Hutton
Blaine, MN

embellished bustlines, but female comic book characters still had a *long* way to go in the area of equality—and so did female comic book writers. **Tina Robbins**, author of the first all-woman comic, *It Ain't Me, Babe*, led the charge in the 1970s, but it would be another twenty-five years before writers like **Gail Simone** (*Birds of Prey*, *Wonder Woman*) and **Devin Grayson** (*Batman: Gotham Knights*, *Nightwing*) would develop the female persona in comic books and graphic novels.

Luckily, it didn't take quite as long for women to break through the fourth wall of the video game industry. In 1979, a gamer named **Roberta Williams** helped to create Mystery House, the first interactive program that used both text *and* graphics, thus birthing the graphic adventure genre; however, the first woman game programmer and designer is credited as **Carol Shaw**, who designed 3-D Tic-Tac-Toe for the Atari 2600. In fact, fangirls have been responsible for a lot of video game firsts: **Dona Bailey** was the first woman to design an arcade game (Centipede), **Amy Briggs** created the first adventure game for girls (Plundered Hearts), and, of course, let's not forget **Doris Self**. At the ripe old age of fifty-eight, Doris was one of the first female competitive gamers and, in 1983, broke the world high-score record for Q*Bert. She even appeared in the movie *The King of Kong: A Fistful of Quarters* and accepted a Q*Bert arcade machine from the Pac-Man world champion Billy Mitchell. Doris passed away in 2006, but she'll always have a high score in our hearts.

> "Every time Gail Simone replies to me on Twitter or Tumblr, my heart skips a beat."
> *Ashly Nagrant*
> *Pittsburgh, PA*

FANGIRL GEEK GODDESSES

It's thanks to the fab fangirls of yesterday that we can truly appreciate where fangirls are today. Check out the latest generation of fangirl geek goddesses who are pwning ass and taking names. Looks like the world is a geek girl's virtual oyster. (No, really. I heard Activision is developing a new RPG that takes place in a giant mollusk.)

BONNIE BURTON, Voted Jedi-Master Fangirl

Ever since migrating to Lucasfilm in 2003 to become an online content developer, Bonnie Burton's name has become synonymous with *Star Wars* and all things Wookiee. Not only does she run the kids section of StarWars.com, but she's also built a lucrative craft career centered on the sci-fi saga and penned two how-to books, *You Can Draw Star Wars* and *The Star Wars Craft Book*. However, there's a lot more to Burton's fangirl glory than her work in the Galactic Empire. She also founded one of the first female-centric pop-culture websites, Grrl.com, and wrote about mean girls—long before Tina Fey, mind you—in her essential read *Girls Against Girls: Why We Are Mean to Each Other and How We Can Change.*

"Self-confidence goes a long way, and I think girls need to realize that at a young age. You have to discover on your own that you're cool in your own right, and you don't have to prove it to everyone."
—editor-writer Bonnie Burton, StarWars.com

ROSARIO DAWSON, Voted Unexpected Trekkie Fangirl

Everyone knows Rosario Dawson is a talented actress, having appeared in cult-classic movies like *Kids*, *Sin City*, and *Grindhouse: Death Proof*, but what everyone *doesn't* know is that Dawson can speak Klingon, reads Neil Gaiman, and is a longtime fan of alt-comic creator Jhonen Vasquez.* In other words, she's a loud and proud fangirl. Her love of comics inspired her to cocreate and cowrite the series *O.C.T.: Occult Crimes Taskforce*, which is about a New York detective who's recruited by a covert team of investigators to combat the supernatural. Dawson's series even scored the endorsement of king fanboy Kevin Smith, who called her "the hottest geek on Earth," and can be found in his comic book store Jay and Silent Bob's Secret Stash in Red Bank, New Jersey.

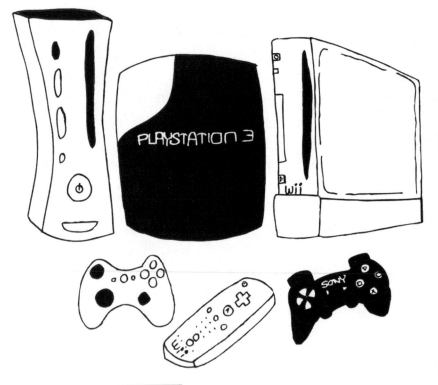

---

* Best known for his satirical style and the comic *Johnny the Homicidal Maniac*.

THE FRAG DOLLS, Voted High-Scoring Fangirls

Formed in 2004 by the video game developer Ubisoft, the Frag Dolls are a group of girl gamers who compete in tournaments, appear at gaming conventions, and, most important, promote fangirl power throughout the gaming realm. All the girls go by aliases—like Fidget, Rhoulette, or Valkyrie*—and while some have retired over the years, the team is always adding new Dolls. They've even started the Frag Doll Cadettes Academy, which is like an industry apprenticeship for aspiring girl gamers. Sadly, there aren't (yet) a *ton* of girl gamers out there, at least when compared to the male presence, so what the Frag Dolls are doing to raise awareness and unite the gaming community is revolutionary.

JOËLLE JONES, Voted Comic Book Artist Extraordinaire Fangirl

If my life were to be made into a comic book,† then I'd definitely want Joëlle Jones to illustrate it. Her style mixes together amplified realism with a touch of cartoonish fantasy and a dash of manga for good measure. This ridiculously talented artist has drawn everything from one-shot comics (like *Dr. Horrible*, based on Joss Whedon's *Dr. Horrible's Sing-Along Blog* web series), romantic graphic novels (like *12 Reasons Why I Love Her*, which is a compendium of twelve vignettes outlining the love trajectory of a young hipster couple, Gwen and Evan), and bewitching teen comedies (like *Spell Checkers*, which writer Jamie S. Rich describes as "*Mean Girls* meets *The Craft*.") Joëlle's talent can only be matched by her productivity, so be sure to check out the "Daily Doodles" on her website, www.joellejones.com.

---

* Not to be confused with the Tom Cruise Nazi movie of the same name.
† I'm game if you are, Dark Horse.

OLIVIA MUNN, Voted Fangirl Geek Goddess Pinup

Much like Antoine Dodson and Spaghetti Cat, Olivia Munn is a cultural phenomenon. When she's not starring in prematurely canceled sitcoms (RIP *Perfect Couples*), the former G4 *Attack of the Show!* cohost acts as a news correspondent for *The Daily Show*. Then, somewhere in between, she manages to appear in movies like *Iron Man 2*, pose for *Maxim* magazine, write books like *Suck It, Wonder Woman: The Misadventures of a Hollywood Geek*, and start her own magazine, *Hey Olivia!* She's basically the fangirl version of Oprah—minus the gazillion dollars, although it's probably only a matter of time. Olivia might get some flack because she flaunts her sexuality (something that's traditionally very *un*-fangirl-like), but that shouldn't tarnish her goddess glow. If anything, it just proves that fangirls come in all different shapes, colors, and cup sizes.

HALL OF FAME: FELICIA DAY

Up until 2007, Felicia Day was your average aspiring actress. She had a smattering of forgettable roles on her résumé (like "Call Girl" in the made-for-TV movie *They Shoot Divas, Don't They* and "Jessie's Friend" in the doctor drama *Strong Medicine*) and spent most of her days waiting for her agent to call. Actually, that's not true. She spent most of her days playing World of Warcraft. So, as a way to keep busy and not completely fall off the acting grid, she decided to produce her own web series called *The Guild*, which centers around Cyd Sherman (aka Codex) and the group of like-minded misfits she meets on- and offline while playing an MMORPG game.

As of August 2010, *The Guild* had over 45 million views on YouTube, generated a limited-series comic book published by Dark Horse, and spawned two viral music videos for "(Do You Wanna Date My) Avatar" and "Game On." In the midst of it all, Felicia Day managed to become a fangirl icon, proving that it's possible for your passion to become your profession. These days, this Comic-Con darling's dance card is pretty packed: she plays the lead in *Red*, the Syfy channel's updated adaptation of *Little Red Riding Hood*; she lends her voice to the video game *Guild Wars 2*; and, of course, there's still talk of her reprising the role of Penny in the long-anticipated sequel to Joss Whedon's musical miniseries *Dr. Horrible's Sing-Along Blog*. In other words, people can't get enough of Day's geeky charm and fangirl wit, and instead of her changing who she is to fit in with Hollywood, Hollywood is willing to carve out a place for her—gamer thumb and all.

DRAWN TOGETHER

Despite popular belief, comic books aren't just a pastime for twenty-three-year-old male assistant managers at GameStop who still live in their parents' basement. Actually, tons of girls are getting in touch with their inner graphic-novel enthusiast, thanks to a few awe-inspiring animated role models.

BUFFY SUMMERS

Buffy Summers has come a long way since her first incarnation as a ditzy blond cheerleader (with an affinity for half-shirts) in the 1992 cult film *Buffy the Vampire Slayer*. Creator Joss Whedon knew she was destined for more than the bargain DVD bin at your local Costco, so he decided to revive the character for his long-running WB series of the same name—and a post-feminist icon was reborn. Throughout her tenure at Sunnydale High School, Buffy constantly struggled with her Slayer status, mostly because it made her classmates look at her like she was a total weirdo. Oh, and let's not even talk about its effect on

"Remember to always be yourself. Unless you suck."
—Buffy Summers, in *Buffy the Vampire Slayer*

her romantic relationships. Thankfully, no matter how badly Buffy was plagued by her powers, she never stopped fighting the vampires, the bad guys, and, of course, the haters who didn't understand her awesomeness. Sure, the show may've ended in 2003, but Buffy continues to kick undead ass in her Dark Horse comic book series.

DARIA

Before you say it, I already know. Technically speaking, Daria Morgendorffer *isn't* a comic book character. (She originally appeared as a supporting character in MTV's *Beavis and Butt-Head* and was then given a spin-off series in 1997.) However, if there's one animated antiheroine who has instilled the importance of satire and intellect in the hearts of fangirls everywhere, it's Daria. To say that Daria is *kinda* sarcastic would be like saying Katie Price's boobs are *kinda* big. However, beneath her pessimistic and misanthropic exterior lies a teenage girl who wants to experience normal teenage girl things—like falling in love, not killing her parents, and getting through high school unscathed. Unfortunately, her higher level of intelligence often makes people nervous and keeps most at arm's length. What fangirl can't relate to that?

"People call me a feminist whenever I express sentiments that differentiate me from a doormat or a prostitute."
—Daria Morgendorffer, in *Daria*

DEATH

Fangirls were first introduced to Death when she appeared in Neil Gaiman's epic series *The Sandman* as Dream's sister. However, she became so popular that Gaiman gave her two solo spin-offs—*Death: The High Cost of Living* and *Death: The Time of Your Life*. It doesn't take a philosophy major to know that Death signifies—wait for it—death. However, unlike previous animated renderings of the Grim Reaper, our DiDi takes the form of a beautifully mysterious goth girl. She's decked out

in all black, rocks some major raccoon eyes, doesn't go any-
where without her ankh necklace, and also happens to worship
Robert Smith. Along with leaving the scythe at home, Death is
oddly Zen and optimistic. She's like that one objective friend
who always manages to put everything in perspective. After all,
no matter how bad you think things are, they could always be
worse. You could be, well, dead.

EMILY THE STRANGE

Emily the Strange might've turned thirty in 2011, but she doesn't
look a day over thirteen—mostly because that's the way artist-
skateboarder Rob Reger created her. (That and the fact that she
never smiles, which means she never has to worry about crow's
feet.) Emily is hardly your typical, happy-go-lucky, *iCarly*-
watching tween who collects Juicy Tube lip gloss and obsesses
over Justin Bieber. Instead, she's a guitar-playing anarchist-
in-training who prefers the company of her cats to that of any
human and spews the kind of cynical philosophy that would
make Dorothy Parker* jealous. Her no-BS POV has earned her
icon status with independent-minded girls (and necromantic
guys) of all ages and resulted in an empire of comic books, clothing, hair dye, soda pop,† sleeved blankets, stackable washer dry-
ers, and sexually suggestive exercise gadgets.‡ Okay, I made up
those last two, but they're not too far off from reality. Beliebers§
should be afraid. *Very* afraid.

"Be your own hero." —Emily the Strange

---

* Sharp-tongued poet, critic, and satirist who rose to literary infamy in the
1920s. Much more on her later!
† This one's actually real. Emily teamed up with Jones Soda to create a line of
limited-edition labels featuring Miss Strange.
‡ Yeah, I'm looking at you, Shake Weight®.
§ The name of Justin Bieber fanatics.

TANK GIRL

In the mid-1990s, Tank Girl was the portrait of grunge-era, twentysomething American angst, which is ironic because the comic's creators were British and the series took place in Australia. Weird. Tank Girl isn't necessarily the *best* role model, what with her relationship with a drug-addled kangaroo, her affinity for public intoxication, and her opposition to wearing shirts and all,* but that's why she's badass. Some *Tank Girl* purists think the series lost major cred when it was made into a movie with Lori Petty as the lead character and Ice-T, dressed in a kangaroo suit,† as her boyfriend Booga. That said, the movie was a humongous, gigantic, jaw-dropping flop. Either way, Tank Girl will always be a postpunk icon because she's everything you wish you had the cojones to be—that is, assuming you subscribe to *Bust* magazine and still listen to a lot of Bikini Kill.

---

* Or bras, for that matter. It takes a certain type of gal to get away with the Wendy O. Williams duct-tape-over-the-nipple style . . . and that gal shouldn't be rocking more than a B cup.

† I think that last part deserves to be repeated. Ice-T . . . dressed in a *mother-effing* kangaroo suit! Take a minute and let that sink in. How was this a flop?

FRENEMIES

To certain people, comics are simply things that appear in the Sunday section of the newspaper—and usually end up lining the bottom of the cat's litter box. These detractors don't understand the difference between *Beetle Bailey* and *Birds of Prey*, and they certainly can't appreciate why anyone would want to keep reading either after they hit puberty. Much of these haters' heckling stems from their own ignorance; they don't comprehend a fangirl's level of passion or admiration toward comics, video games, anime, sci-fi, or what have you, because they've never given those things a shot. They'd deserve our sympathy if they weren't so dead set on making us feel like overweight, acne-ridden, socially inept losers. After all, not all fanboys look like Comic Book Guy from *The Simpsons** and not all fangirls rock the Princess Leia buns on a daily basis. And that's why these fangirl frenemies are the worst . . . *ever.*

✔ Self-proclaimed Twilighters who either (1) didn't notice that Bryce Dallas Howard replaced the character of Victoria, who was previously played by Rachelle Lefèvre, in *Twilight: Eclipse*, or (2) did notice but weren't *totally* outraged by it.

✔ Anyone who gets *Star Wars* confused with *Star Trek.*†

✔ Literary snobs who think comic books aren't "real" books.

✔ The members of Westboro Baptist Church who picketed Comic-Con because they think participating in cosplay makes you a devil worshipper.

✔ Avid TV watchers who've never seen an episode of *Lost* but insist that they totally "get it."

---

* That is, old, fat, balding, lonely, chaste, still obsessed with Pogs . . . Need I go on?
† How is that even humanly possible?

✔ Athletes.

✔ Obsessive Twitter users who only tweet in LOLcat speak (e.g., "me hungryz, nom nom nom" or "I needz halp wit mai body ish00z").

✔ Houseguests who see your Sony PlayStation and ask if you live with a ten-year-old.

THE *TWILIGHT* ZONE

For the nine of you* out there who've been living in a cave for the past few years, Stephenie Meyer is the hugely successful, wildly worshipped author of the *Twilight* books. As the legend goes, Meyer had no previous writing experience before penning the first *Twilight* novel, but after the idea came to her in a dream—that of a human girl falling in love with a devastatingly handsome vampire—she sat down

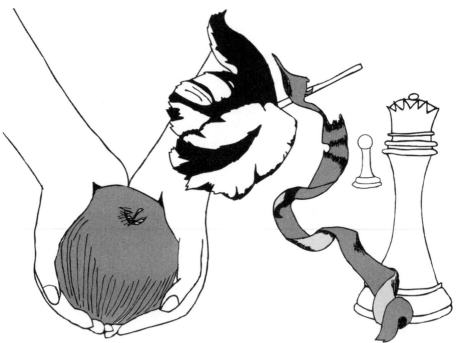

* I realize I might even be overestimating.

and diligently wrote out her fantasy, and three months later she had herself a book. To date, the series has sold over 100 million copies worldwide and has been translated into thirty-seven languages.

Love it or hate it, *Twilight* is here to stay—and so is RPattz, thankfully. (*Exhales sigh of relief.*) However, is this vampire saga a true fangirl delicacy or just an excuse for minivan-driving soccer moms to drool over shirtless pieces of fresh male meat without being called Mary Kay Letourneau? Time to get to the bottom of this once and for all. What kind of Twi-mom behavior is appropriate, and what kind should prompt a call to Chris Hansen?

If you're older than twenty-two, 72% of ΓΓΓ members think you're too old to worship *Twilight*.

Appropriate: To watch the *Twilight* movies multiple times.
Inappropriate: To watch the *Twilight* movies multiple times but then replay all the half-naked Taylor Lautner shots in slow motion.

Appropriate: To invite like-minded Twi-moms over to your residence *once in a while* to partake in *Twilight*-themed celebrations.
Inappropriate: To throw weekly *Twilight*-inspired gatherings, complete with personalized jewelry and food, where you speculate about which sexual encounter between Bella and Edward resulted in the conception of Renesmee.

Appropriate: To relive your teenage years while reading *Twilight*.
Inappropriate: To condemn your husband or boyfriend because you two will never have the kind of eternal love Bella and Edward have.

Appropriate: If you ever find yourself in the company of a *Twilight* cast member, to ask said actor or actress to take a picture with you.

Inappropriate: If you ever find yourself in the company of a *Twilight* cast member, to ask said actor or actress to autograph your underwear. Bonus points if your teenage son or daughter is present.

Appropriate: To be sad when the *Twilight* films end production.
Inappropriate: To say you suffer from PTSD (post-*Twilight* stress disorder) and insist that you need to seek immediate medical attention.

Appropriate: To be frustrated with your teenage son or daughter, who doesn't understand your obsession with *Twilight*.
Inappropriate: To be frustrated with your teenage son or daughter, who doesn't understand your obsession with *Twilight* . . . and then daydream about all the different ways you could kill them in their sleep.

GEEK LOVE

It's not easy to win the heart of a fangirl geek. After all, she's content to immerse herself in other realms (e.g., World of Warcraft's Azeroth or *The Lord of the Rings*' Middle Earth); it takes an über-special guy to bring her back into the arms of reality. A common interest in video games, comic books, and science fiction is a must, in addition to an understanding of proper grammar and sentence structure. After all, a guy who speaks only in emoticons is a guy you want *nothing* to do with. (Trust me.) Also, a fangirl geek paramour doesn't get caught up in style trends or fashion fads, although he should know how to dress himself in a way that doesn't make him look like an overgrown toddler.[*]

However, perhaps the most important part of a blossoming relationship with a fangirl is communication. One must be able to communicate across a bevy of different platforms—like email, Instant Messenger, text, BBM, FaceTime, Skype, etc. But remember: it's one thing to be tech-savvy and quite another to be tech-dependent. Nothing ruins romance more than a guy who can't converse unless he's behind a screen. Face-to-face contact is essential, if for no other reason than to prove whether your future boyfriend is who he says he is . . . *and* that he can string more than five words together without having to press SEND.

"I wish I could freeze this moment, right here, right now, and live in it forever."
—Peeta Mellark, in *Catching Fire* by Suzanne Collins

---

[*] This means no sweatpants. Ever.

THE PERFECT MATCH FOR A FANGIRL GEEK . . .

❏ Looks like Neil Gaiman, Nathan Fillion, or David Tennant.

❏ Knows that HBO's *True Blood* is actually based on *The Southern Vampire Mysteries/The Sookie Stackhouse Chronicles* by Charlaine Harris.

❏ Is proud to be a beta male.*

❏ Would rather stay home and play Apples to Apples than go see a Jack Johnson wannabe butcher songs at the corner bar's open-mic night.

❏ Is assessed as an ENTP personality type,† according to the Myers-Briggs Type Indicator.‡

❏ Daydreams about how to survive a possible zombie apocalypse . . . or has multiple action plans in place and ready to go if and when the day comes.

❏ Owns *at least* one T-shirt with the Superman, Batman, or Green Lantern emblem on it.

❏ Isn't afraid to step out from behind the computer screen and leave the house.§

---

* Beta males are the opposite of alpha males. They tend to be intelligent, quiet, and nonconfrontational. They also aren't afraid to let their lady shine in the spotlight. *Swoon.*

† The abbreviation for "extraversion, intuition, thinking, perception."

‡ More important, he knows what the Myers-Briggs Type Indicator *is*. Just so we're clear, it's a psychological questionnaire used to measure how people look at the world and how they make their decisions. There. Now we're all on the same page.

§ Unless homeboy has a sun allergy.

COMIC-CON: WHERE THE FANBOYS ARE

Patti Stanger, otherwise known as the Millionaire Match-maker, might not be the smartest woman on the planet,* but she's definitely right about one thing: fan conventions are a great place to meet your geek soulmate. For the fangirl geek, specifically, that means Comic-Con International, the annual horror, anime, manga, animation, toy, video game, comic book, and fantasy novel convention. When the convention started in 1970, it only drew about 300 people. Fast-forward forty years, and now nearly 140,000 geeks descend upon San Diego every June, just like the 501st Legion descended upon the Jedi Temple in *The Clone Wars* . . . or something. Sure, not *all* of them are genuine fanboys, but the odds are definitely in a fangirl's favor. Just make sure to avoid anyone who: (1) doesn't wear deodorant, (2) gives out free hugs, (3) asks if you want to see the size of his sword, and/or (4) is pushing around a grocery cart full of Ewoks.

If you can't tote your cookies to San Diego, try to check out the following conventions happening in your neck of the woods:

ALTERNATIVE PRESS EXPO (comic-con.org/ape)
Location: San Francisco, CA
Genre: comic books

ANIME EXPO (anime-expo.org)
Location: Los Angeles, CA
Genre: anime

---

* Although she *might* be one of the most annoying.

BLIZZCON (blizzcon.com)
Location: Anaheim, CA
Genre: World of Warcraft, StarCraft

CELEBRATION (starwarscelebration.com)
Location: Various
Genre: all things *Star Wars*

DRAGON*CON (dragoncon.org)
Location: Atlanta, GA
Genre: costuming, fantasy, gaming, science fiction

GEEK GIRL CON (geekgirlcon.com)
Location: Seattle, WA
Genre: science fiction, comics, gaming, geek culture

LEAKYCON (leakycon.com)
Location: Orlando, FL
Genre: all things *Harry Potter*

MEGACON (megaconvention.com)
Location: Orlando, FL
Genre: comic books, science fiction, fantasy, anime

THE SLAYAGE CONFERENCE ON THE WHEDONVERSES
(slayageonline.com)
Location: Various
Genre: all things Joss Whedon

WONDERCON (comic-con.org/wc)
Location: San Francisco, CA
Genre: comic books, science fiction, and movies

WYRDCON (wyrdcon.com)
Location: Costa Mesa, CA
Genre: LARP

REQUIRED READING

The Amazing Adventures of Kavalier & Clay by Michael
  Chabon
The Chronicles of Narnia series by C. S. Lewis
The *Harry Potter* series by J. K. Rowling
The Hunger Games series by Suzanne Collins
Inkheart trilogy by Cornelia Funke
Lamb: The Gospel According to Biff, Christ's Childhood Pal
  by Christopher Moore
The Lord of the Rings series by J. R. R. Tolkien
Matilda by Roald Dahl
The Mortal Instruments series by Cassandra Clay
Percy Jackson and the Olympians series by Rick Riordan
Star Wars Craft Book by Bonnie Burton
Stardust by Neil Gaiman
The *Twilight* saga by Stephenie Meyer
True Blood boxed set (Books 1–8) by Charlaine Harris
Watchmen by Alan Moore
A Wrinkle in Time quintet boxed set by Madeleine L'Engle

COMIC BOOK GENIUS

Birds of Prey by Gail Simone
Ghost World by Daniel Clowes
The Guild by Felicia Day

Indoor Voice by Jillian Tamaki

Jack Staff by Paul Grist

Kick-Ass by Mark Millar

Lucky by Gabrielle Bell

Make Me a Woman by Vanessa Davis

Persepolis by Marjane Satrapi

The Sandman by Neil Gaiman

Scott Pilgrim vs. the World by Bryan Lee O'Malley

The Umbrella Academy by Gerard Way

WEB BOOKMARKS

AllThingsFangirl.com

Comic-Con.org

FantasticFangirls.org

FragDolls.com

GeekGirlsNetwork.com

Geek.MTV.com

Geektress.com

GeekWeek.com

Grrl.com

NeilGaiman.com

She-Geeks.com

SplashPage.MTV.com

TheDiscriminatingFangirl.com

TheMarySue.com

TheNerdyBird.com

WilWeaton.Typepad.com

MUST-SEE MOVIES

Avatar

The *Back to the Future* trilogy

A Clockwork Orange

E. T. Extra Terrestrial

Fanboys

Goonies

The *Harry Potter* films

The Lord of the Rings trilogy

The *Millennium* trilogy: *The Girl with the Dragon Tattoo,*
  The Girl Who Played with Fire, The Girl Who Kicked the
  Hornet's Nest

The NeverEnding Story

The *Star Wars* films

Titanic

The Twilight Saga

Two

LITERARY GEEK

Sure, you've read *Sex, Drugs, and Cocoa Puffs: A Low Culture Manifesto*, but can you name any of Chuck Klosterman's other books*—the ones that *aren't* sold at Urban Outfitters? Power down your iPad, grab a Magic Marker, and put your literary geek girl skills to the test!

1. *The Catcher in the Rye*, J. D. Salinger's novel about teenage angst, seems to be required reading for disaffected youth—and serial killers—everywhere. That's right. The

---

* For example, *Fargo Rock City: A Heavy Metal Odyssey in Rural North Dakota* (2001), *Killing Yourself to Live: 85% of a True Story* (2005), *Chuck Klosterman IV: A Decade of Curious People and Dangerous Ideas* (2006), *Downtown Owl: A Novel* (2008), *Eating the Dinosaur* (2009), and *HYPERtheticals: 50 Questions for Insane Conversations* (2010).

book has been linked to various high-profile killers, including:

A. Charles Manson, who was responsible for the Tate-LaBianca murders.

B. "Wait a sec? Why would a book about baseball make people kill each other?" (*Scratches head and goes back to watching* Burn Notice *on the USA Network.*)

C. Mark David Chapman,* who was responsible for the murder of John Lennon.

2. In 2005, an eighteen-year-old literary prodigy landed a two-book deal (allegedly worth $500,000) with Little, Brown. This debut novel, *How Opal Mehta Got Kissed, Got Wild, and Got a Life,* was released a year later during the author's freshman year at Harvard. Following the book's release, the *Harvard Crimson* reported that *Opal Mehta* appeared to have been plagiarized from numerous young-adult novels and the book was eventually pulled from the shelves. The name of said author is:

A. James Frey.

B. Lilian Jackson Braun.

C. Kaavya Viswanathan.

3. Which of the following is *not* the name of an ebook device?

A. NOOK.

B. Kindle.

C. Parchment Paper.

4. Irish author James Joyce considered *Finnegans Wake* to be his finest work. Comprising seventeen chapters and divided into four books, the tome clocks in at almost seven hundred pages

---

* Other murderous fans include John Hinckley, who attempted to assassinate Ronald Reagan in 1981, and Robert John Bardo, who was carrying the book when he killed actress Rebecca Schaeffer.

and is widely considered one of the most difficult reads in modern literature. How long did it take Joyce to complete his self-proclaimed masterpiece?

A. Seven years.

B. "Who cares? I'd rather read the phone book than this waste of trees."

C. Seventeen years. And did you know the first sentence is also the last sentence?

5. *Vogue* is to Condé Nast as *The Believer* is to:

A. The Hearst Corporation.

B. Ozzy Osbourne.

C. McSweeney's*.

ANSWER KEY

Mostly As: While it's obvious that you read more than just the year-old issues of *Us Weekly* in the doctor's office, you're going to have to up the literary ante if you want to be considered a true literary geek girl.

Mostly Bs: Sorry, Lauren Conrad, but just because you have a good ghostwriter doesn't mean you'll ever win a Pulitzer.† In fact, you should probably stick to designing clothes for Kohl's and dating Z-list celebrities before you hurt that pretty little brain of yours.

Mostly Cs: High fives, high priestess. You're a bona fide literary geek girl!

---

* McSweeney's is a publishing house founded by editor Dave Eggers. It's responsible for publishing a quarterly literary journal and DVD magazine, a critically acclaimed website, and the monthly magazine *The Believer*.

† Before you ask, LC, no, they don't sell Pulitzers at Intermix.

CHARACTER SKETCH

Ever since the 1960s, upon the urging of Dr. T. Berry Brazelton and the all-knowing Dr. Spock,* mothers have been encouraged to read to their children at a very early age. For toddlers and preschoolers who relish this early diet of literacy, libraries become a second home, story hour is never long enough, and parents can't finish a book without hearing a little voice beg, "Again . . . *again.*" For most literary geek girls, it's at this age that they discover their passion for reading. Whether it's *Harold and the Purple Crayon* or *Strega Nona*, books provide the budding literary she-geek with a glimpse into an all-new world of magic and make-believe—and once she visits, she immediately wants to apply for full-time citizenship.

> "We tell ourselves stories in order to live." —author Joan Didion, in *The White Album*

While some children spend their summers sweating on community sports teams or learning Indigo Girls songs at sleep-away camp, our beloved bookworms are more interested in joining their local library's summer reading program, completing twenty-five books during vacation, and earning a certificate of recognition signed by their city's mayor. (Plus, that Sony Bloggie Touch the library is giving away to the person who logs the most hours reading *isn't* the worst incentive, either. It'll come in handy for that book review YouTube channel she's been thinking about starting!) When school starts back up again, her friends will inevitably show off their tan lines and pony bead friendship bracelets, and our geek girl will politely oblige by oohing and aahing accordingly. But secretly she's bursting with pride over her summer's battle scars—the numerous paper cuts she got while feverishly turning the pages of all seven *Harry Potter* books.

---

* A pediatrician turned author who revolutionized child rearing with his book *The Common Sense Book of Baby and Child Care*, which asserted that babies should be raised with affection over discipline. Not to be confused with Mr. Spock, the Vulcan lieutenant commander on *Star Trek*.

Because the literary geek girl spends most of her free time with her nose in a book, it's hard for her to keep an eye on popular fashion and music trends. Her "look," if you can even call it that, is oftentimes simple and understated: jeans (probably secondhand or from Target), a T-shirt (possibly boasting a slogan like I ~~BRAKE~~ GO BROKE AT BOOKSTORES), and some kind of sneakers (anything ranging from Converse to Keds to Asics). Oftentimes, her appearance will evolve over the years into something that resembles a sexy librarian* or eccentric schoolmarm.† When it comes to tunage, don't be surprised if our she-geek is super into classical music—or bands that are influenced by nineteenth-century German poets and effortlessly work fifty-cent words like "anathema" into their lyrics.‡ It might seem dated and dorky to some, but much like literature itself, this type of orchestral composition is completely open to interpretation and encourages the use of imagination—something that the literary geek girl obviously thrives upon.

All in all, if you really want to crawl into the mind of this particular geek girl and understand what makes her tick, look no further than the whimsical—yet completely apropos—theme song of *Reading Rainbow*, the famed PBS children's television series starring LeVar Burton. It really explains it all.

> "[The typical literary geek girl] wants to read the book before she sees the movie. Everyone she meets is a 'character,' and everything she lives is a 'story.'"
> *Bree McGuire*
> *Los Angeles, CA*

I can be anything
Take a look
It's in a book
A Reading Rainbow

---

* Like Joan Holloway on *Mad Men*.
† Like Susan Boyle.
‡ E.g., Rainer Maria, a 1990s-era emo band that was named after nineteenth-century German poet Rainer Maria Rilke and wrote a song called "Artificial Light," which not only featured the word "anathema" but also the word "atrophy." Talk about an impressive double-word score!

BORN TO READ

Kids say the darnedest things—especially when they're inspired by precocious literary heroines like the ones below. Take a seat on the magic carpet and let's revisit some of the characters who made us bookworms first fall in love with reading and writing.

RAMONA QUIMBY

from *Beezus and Ramona* by Beverly Cleary
A lot of people view Ramona as the typical younger sister—loud, annoying, and nosy; however, beneath her pestlike exterior lies an inquisitive little girl who has an insatiable interest in the world and the well-being of the people around her. Her wild imagination, fierce independence, and unwavering curiosity make her a role model for geek girls of all ages.

MARGARET SIMON

from *Are You There, God? It's Me, Margaret* by Judy Blume
It's not easy being a preteen girl, especially when you're in the throes of puberty, aren't sure what you believe in, and have to deal with belted sanitary napkins. (*Shudder.*) Thankfully, for all us girls who feel like we don't fit in, we've got an ally—and an advocate—in Margaret.

NANCY DREW

from the *Nancy Drew* mysteries by Carolyn Keene
When Nancy Drew first started solving crimes back in the 1930s, she was a groundbreaking character. She challenged the idea that girls were weak, fearful, and passive. After all, you have to be pretty fearless and outspoken to be a teenage

detective. Today, Nancy might be well into her eighties, but she doesn't look a day over sixteen. How *does* she do it?

MARY LENNOX

from *The Secret Garden* by Frances Hodgson Burnett

With the help of a couple of gardening tools and the faith of a few newfound friends, Mary Lennox miraculously transforms from a sickly spoiled brat whom everyone avoids to an empathetic do-gooder whose spirit is intoxicating. Plus, in addition to going through her own physical and emotional transformation, she helps to convince all around her that anything is possible as long as you believe.

MATILDA WORMWOOD

from *Matilda* by Roald Dahl

Sensitive, intelligent, kind, and driven, book-loving Matilda would be the perfect child for anyone except the loathsome Wormwoods, who see their gifted daughter as nothing but a bother and a pain. (Someone call *Nanny 911!*) Unbeknownst to her parents, not only is Matilda somewhat of a prodigy, but she also possesses psychokinetic powers, which eventually come in handy when so many of the adults around her attempt to derail her quest for an education. Always remember, kids: knowledge is power, but karma is a b*tch.

HARRIET M. WELSCH

from *Harriet the Spy* by Louise Fitzhugh

Harriet M. Welsch is a little girl with big dreams. In addition to being a writer, she also wants to be a spy, so in order to hone her skills, she scribbles all her thoughts and observations in her trusty notebook. Unfortunately, not all her insights are appreciated, especially when her friends discover her notebook and are less than pleased with her surveillance. Honesty is important, but so is tact. Luckily for Harriet, the lesson isn't lost on her and she manages to make good on a bad situation. The moral of the story? If you don't have anything nice to say, don't say it—or at least don't say it in a notebook that you lose easily.

GEEK MYTHOLOGY

Book-smart beauties are a captivating breed and they date as far back as the year 1000, when a Japanese lady-in-waiting named Murasaki Shikibu wrote *The Tale of Genji,* which is credited with being the earliest known novel in history.* Plenty of literary ladies contributed works of poetry and fiction in the centuries to follow, but it wasn't until the early 1900s that a real movement seemed to take shape, thanks in part to the work of one particular witty deity: **Dorothy Parker.** Anything Parker wrote—whether it was theatrical criticism, short stories, or simple one-liners—was whip-smart, keenly observant, and often hilarious. She was a mistress of wordplay, and her couplets are just as applicable today as they were during her era. (Example: "You can lead a horticulture, but you can't make her think.")

> "Every woman is a rebel, and usually in wild revolt against herself."
> —author Oscar Wilde

While Parker was busy making quips, fellow female writers of the time were writing beautiful sonnets, like those of poet **Edna St. Vincent Millay,** or third-person autobiographies, like *The Autobiography of Alice B. Toklas* by **Gertrude Stein.** Others, such as **Virginia Woolf,** took a new approach to fiction in the 1920s by exploring characters' emotions and psychological motives in critically acclaimed books like *Orlando* and *Between the Acts.* In 1943, a Russian philosopher and author by the name of **Ayn Rand** released the first of her revolutionary novels, *The Fountainhead,* which was about an architect who seeks fame without sacrificing his integrity and acted as an extension of her Objectivism philosophy.† She followed up with *Atlas Shrugged* in 1957, and its publication solidified her as both a

---

* The first female poet wouldn't be published for another six centuries. Her name was Anne Bradstreet and her collection of poetry, *The Tenth Muse Lately Sprung Up in America,* was released in England in 1650.

† The Ayn Rand Institute defines objectivism as "the concept of man as a heroic being, with his own happiness as the moral purpose of his life, with productive achievement as his noblest activity, and reason as his only absolute."

feminist and an individualist icon. Then, in 1949, **Simone de Beauvoir**—a French existentialist and the well-publicized lover of fellow philosopher Jean-Paul Sartre*—released *The Second Sex*, which foreshadowed the ethos and many of the ideas that became the foundation of the feminist movement (e.g., gender identity, patriarchy, the power of sexuality).

> "A woman must have money and a room of her own if she is to write fiction." —author Virginia Woolf, in *A Room of One's Own*

The 1960s was an exciting time in the world of publishing, especially for literary geek goddesses. Poet and civil rights activist **Maya Angelou** invented a new literary genre—autobiographical fiction—with the release of her memoiresque debut book, *I Know Why the Caged Bird Sings*, which tackled then-taboo topics like racism and teen pregnancy. Future editor in chief of *Cosmopolitan* magazine **Helen Gurley Brown** dared to declare women sexual creatures in her groundbreaking how-to book *Sex and the Single Girl*, while **Harper Lee**, close friend and confidante of Truman Capote, went on to win a Pulitzer Prize for her one and only novel, *To Kill a Mockingbird*, which depicted racial injustice in Lee's hometown of Monroeville, Alabama, in the years following the Great Depression. Speaking of depression, no author captured the angst and anxiety of young womanhood quite like **Sylvia Plath**. Her 1963 semiautobiographical novel *The Bell Jar* chronicled the life of Esther Greenwood, an aspiring magazine writer in her twenties who struggles with life in New York City while battling her own inner demons. Unfortunately, Plath succumbed to her own crippling depression and committed suicide that same year.

> "The book to read is not the one which thinks for you, but the one which makes you think." —author Harper Lee

Around the same time, two other literary luminaries were just taking off. First came **Betty Friedan** and her groundbreaking book *The Feminine Mystique*, which directly inspired the second-wave feminist movement of the mid-1960s. Before Friedan, the word "sexism" didn't exist and the issue of gender equality was basically a moot point. However, after women—and certain

---

* The two made polyamory sexy waaaay before *Big Love*. Take that, HBO.

"evolved" men—got their hands on the book and realized that it was okay for women to have their own identity outside of housewife and mother, nothing was ever the same in the household, the workplace, and society as a whole. A decade later, newcomer **Erica Jong** would publish her novel *Fear of Flying*, which followed a woman in her twenties on her quest to discover her sexual and intellectual identity. Taking cues from Gurley Brown and Friedan, Jong blended together salacious storylines with sexual politics and the result resonated with readers—especially women who thought steamy hookups were reserved for the pages of Harlequin novels.

During the 1980s and early 1990s, there was plenty of heavy petting in the literary realm thanks to sweeping romance novels by **Danielle Steele, Judith Krantz,** and **Jackie Collins.** However, if you preferred to read books that *didn't* have Fabio on the cover, there's a good chance you were attracted to the sentimental musings of columnist-turned-novelist **Anna Quindlen** or the suspenseful narratives of **Joyce Carol Oates.** Then, of course, there was **Elizabeth Wurtzel** and her tome *Prozac Nation: Young and Depressed in America: A Memoir.* With her grungy good looks and enviable breeding, Wurtzel became a poster girl for Gen X apathy and upper-middle-class angst. Quite the polarizing figure, she was condemned by critics for being a self-indulgent name-dropper, while fans commended her for sharing her personal struggle with depression and mental illness. Tomato, tomahto.

Wurtzel ended up leaving publishing to practice law, which left the door wide open for other quirky, indie-lectual authoresses like **Miranda July** and **Jennifer Egan** to mosey on in. Though she's also an accomplished artist and filmmaker, July's area of literary expertise is short fiction, and her book *No One Belongs Here More Than You: Stories* is filled with verbal vignettes that are both confessional and contemplative. Egan, on the other hand, rose through the literary ranks with her inventive 2001 novel *Look at Me*, which examines the fashion industry through the eyes of Charlotte, a model whose face is disfigured after a car crash and who attempts to find substance in a business based around style. Her recent work *A Visit from the Goon Squad** is another example of Egan's unorthodox approach to fiction, which involves nonlinear chronologies, shifting character perspectives, and the use of PowerPoint presentations as a means of storytelling. I can't wait to see who emerges in the next generation of female wordsmiths because, as Natasha Bedingfield says, "the rest is still unwritten."

> "I try to write parts for women that are as complicated and interesting as women actually are." —author Nora Ephron

LITERARY GEEK GODDESSES

Our literary geek girl history is chock-full of the work of female poets, novelists, satirists, essayists, philosophers, and all-around bitchin' bookworms. Without their unprecedented wordsmith skills, we'd probably still be throwing Tupperware parties instead of hosting book club meetings. (Perish the thought!) Let's check out the latest crop of literary geek goddesses who have the write stuff.

SLOANE CROSLEY, Voted Witty Wordsmith Literary Geek

If it weren't for being locked out of two different Manhattan apartments twice in the same day, Sloane Crosley might not have written her debut collection of essays *I Was Told There'd Be Cake*. (Big ups to ineptitude!) Inspired by the tragic hilarity of that singular experience, Crosley decided to scribble it

---

* The book was awarded the 2011 Pulitzer Prize for fiction. In an interview with the *Wall Street Journal*, Egan advised young female writers to "shoot high and not cower."

up and send it to some friends for poops and giggles. An editor at the *Village Voice* was included in the mailing, and he thought it would make a great piece for the paper. The rest is, well, history. Crosley is kind of like a hipster Carrie Bradshaw minus the six-inch Christian Louboutin heels. Sure, her essays examine love and relationships, but she approaches both with self-deprecating banter and matter-of-fact realism that are utterly refreshing. Her latest book, *How Did You Get This Number*, examines all the major—and minor—life experiences a gal faces on the road from blissful young womanhood to dreaded, gray-haired adulthood. In other words, it's required reading for any literary geek girl in the throes of a quarter-life crisis—or facing down the barrel of any milestone birthday.

"Life starts out with everyone clapping when you take a poo and goes downhill from there." —author Sloane Crosley, in *I Was Told There'd Be Cake*

LORRIE MOORE, Voted Too Funny For Her Own Good Literary Geek

Some women seek therapy from psychologists; others seek help from Lorrie Moore. Specifically, they seek *Self-Help*, Moore's wildly successful and insightful collection of short stories about love, loss, and ambition. Moore seemed destined for literary greatness after winning *Seventeen* magazine's young fiction contest at nineteen, but it wasn't until she graduated with a master's in fine arts from Cornell University that she secured an agent and published her first collection of stories, the majority of which were from her master's thesis, at age twenty-six. Since then, she has written novels (*Who Will Run the Frog Hospital?*), children's books (*The Forgotten Helper*), and more collections of short stories (*Birds of America*), all of which bear her signature wit, cynicism, and the kind of real-life observations that hit a little *too* close to home. Moore is currently the Delmore Schwartz Professor in the Humanities at the University of Wisconsin–Madison.

"All the world's a stage we're going through." —author Lorrie Moore, in *Anagrams*

ZADIE SMITH, Voted Humanist Realist Literary Geek

Though Zadie Smith has said in numerous interviews that she feels safer writing nonfiction,* her novels *White Teeth* and *On Beauty* have established her as one of the top young writers of the past twenty years. Known for her realist writing style, Smith tends to examine racial, religious, and cultural differences in her books, but she never sacrifices plot or character development for the sake of inflated conflict—something that many postmodernist writers often do. Smith's last literary effort was 2007's *The Book of Other People*, which is a collection of short stories she se-

---

* While appearing at the 2010 *New Yorker* Festival alongside Michael Chabon, Smith said, "It makes me feel really anxious writing fiction. I think it's because there's some sort of instinct in me that I want to be right, and in fiction, you can never be right. Even worse than that, for me, is I don't know what I want to write in fiction. I don't know what would be best. When I'm writing nonfiction, I feel that the facts are such a fantastic anchor."

lected and edited by authors like Nick Hornby, Miranda July, Jonathan Lethem, and Smith herself. In 2010, Smith became a tenured professor of fiction at New York University. Budding postmodernists should enroll in one of her courses ASAP!

SARAH VOWELL, Voted History Buff Literary Geek

Sarcastic, loquacious, and obscenely intelligent, Sarah Vowell has a voice that is unforgettable on and off the page. Not only can she be credited with making American history interesting in books like *Assassination Vacation* and *The Partly Cloudy Patriot*, but she also lends her signature snark to the character of Violet Parr, a teenage superhero misanthrope in Disney Pixar's *The Incredibles*. Though some call her a social observer, Vowell prefers to be called a writer, plain and simple. Whether she's producing a radio segment about Rosa Parks for *This American Life* or writing about the Americanization of Hawaii, Vowell manages to make otherwise complicated political topics relatable and, more important, understandable to someone without a PhD in history.

J. K. ROWLING, Voted Rags-To-Wizardry-Riches Literary Geek

Up until 1995, Joanne Rowling* was simply a depressed single mom, recovering from a divorce and trying to put the pieces of her life back together. She was living on government assis-

---

* With the release of the first *Harry Potter* book, Rowling started using the pen name J. K. Rowling. During an appearance on *Oprah*, Rowling said, "My British publisher thought '[*Harry Potter*] is a book that will appeal to boys,' but they didn't want the boys to know a woman had written it." Hence the initials because, you know, women writers have cooties and stuff.

tance in Edinburgh, Scotland, and attempting to get her teaching certificate when something magical happened—*literally*. She was inexplicably struck with the idea for a book about wizards and began writing immediately. Over the next ten years, she created the enchanting world of Harry Potter, penned a series of books filled with Harry's adventures at Hogwarts School of Witchcraft and Wizardry, sold over 400 million books in said series, and became a billionaire.* Not only is Rowling's own story inspiring, but the one she crafted in the pages of *Harry Potter* is something that has inspired, entertained, and united generations of book lovers all over the world. As Dumbledore says in *Harry Potter and the Prisoner of Azkaban*, "Happiness can be found, even in the darkest of times, if one only remembers to turn on the light."

HALL OF FAME: JANE AUSTEN

Whether they call themselves Austenologists or Janeites,† it's hard to find a literary geek girl who doesn't adore the work of Jane Austen. Known for her sharp wit and sentimental irony, Austen basically invented the genre of chick lit with the posthumous success of period novels like *Sense and Sensibility* and *Mansfield Park*. Many also view Austen as one of the earliest first-wave feminists because she dared to write about then-unmentionable topics like gender and class and continued to question whether a woman could truly be both independent *and* married—a debate that still carries on today.

When Jane Austen first published her novel *Sense and Sensibility* in 1811, the London literary community basically

---

* She's actually the first billionaire author and, according to *Forbes*, the 937th richest person in the world.
† Hardcore Jane Austen fans.

yawned. Not only were the themes of Austen's books far too progressive for the time, but also it was downright uncommon and unseemly for women to pursue a career in publishing.[*] Much like her beloved characters, Austen wasn't interested in conforming to traditional literary standards—or the expectations of the aristocracy, for that matter. She wanted to write about *real* women, complete with flaws, intellect, and sass to spare. That's why her fictional creations—like Emma Woodhouse from *Emma* and Elizabeth Bennet from *Pride and Prejudice*—remain feminist icons to this day and continue to be adapted and updated, though never outdone. How many authors can say the same?

FRENEMIES

Sure, there are lots of literary ladies who are total inspirations to the next generation of aspiring writers and bibliophiles, but today's crop of well-read she-geeks isn't without its fair share of detractors. So before you power up that shiny new e-reader, make sure these frenemies aren't throwing any haterade your way.

- ✔ Braggarts who donate to NPR just so they can say they donate to NPR and like to boast about it any chance they get.
- ✔ Anyone who cheated their way through high school and college English literature classes by relying solely on CliffsNotes.
- ✔ Members of the illiterati.
- ✔ Simpletons who are only familiar with the term "wordplay" because it's the name of a Jason Mraz song.
- ✔ Commuters who appear to be reading *Anna Karenina*

---

[*] For this reason, Jane Austen originally published anonymously.

on the subway when, in fact, they're using Leo Tolstoy's book jacket to hide *Safe Haven* by Nicholas Sparks.

✔ Fist-pumping juiceheads and guidettes who pre-ordered *A Shore Thing* by Nicole "Snooki" Polizzi or *Here's the Situation: A Guide to Creeping on Chicks, Avoiding Grenades, and Getting In Your GTL on the Jersey Shore* by Mike "The Situation" Sorrentino.

✔ Twi-hards who argue that Stephenie Meyer is *actually* a good writer.

✔ Bloggers who don't know the difference between "their" and "they're," "your" and "you're," "its" and "it's," "whose" and "who's," and "effect" and "affect."

GEEK LOVE

When it comes to dating, literary geek girls are a picky bunch. They're not going to be impressed by suitors who refer to whatever men's magazines happen to be piled on the top of their toilet tank as "reading material." They aren't going to swoon over dudes who've never heard of David Sedaris. And they definitely wouldn't be interested in a guy who can't name at least one of Shakespeare's plays. In other words, our literary she-geeks are looking for someone with a little more intelligence—and a lot more ingenuity.

At the very least, any guy who's going to win the heart of a literary geek girl *has* to be literate. It's kind of a prerequisite. (Bonus points if he owns a bookshelf. *Super* extra points if said bookshelf is stacked with actual books.) Literary geek girls possess a thirst for knowledge that can never be quenched, which is why many well-read females tend to fall for well-educated men who like to discuss, debate, and spar. Not only do such activities get the brain juices flowing, but they're also a form of flirty foreplay. Our darling bookworms should be mindful, however, of highly intelligent men who are more interested in hearing themselves talk than listening to what you have to say. Intelligence is sexy—but being an argumentative know-it-all is a major turnoff.

THE PERFECT MATCH FOR A LITERARY GEEK GIRL . . .

❏ Looks like Dave Eggers, Jonathan Safran Foer, or James Franco.

❏ Owns at least one corduroy blazer with elbow patches.

❏ Isn't ashamed to admit he has a profile on Alikewise.com.*

❏ Only reads one book at a time and thinks someone who's "in the middle" of numerous titles displays commitment issues.

❏ Can name three publishing houses off the top of his head.†

❏ Always reads the book before seeing the movie based on it—and *almost* always thinks the book is better.

❏ Instead of setting an alarm, prefers to wake up to the soothing sound of Steve Inskeep's and Renee Montagne's voices on NPR's *Morning Edition*.

❏ Is obsessed with Words with Friends on his iPhone.

---

* An online dating site for bookworm singles.
† Including but not limited to HarperCollins, Random House, Simon & Schuster, Macmillan, Houghton Mifflin Harcourt, Harlequin, Scholastic, and Penguin.

BOOK CLUBS ARE THE NEW BLACK

Getting together a gaggle of girl geeks for a round of snacks and literary chitchat is a totally awesome idea. But how can you make sure your book club's got it going on? Here are some novel ideas sure to spark hot dish and even hotter discussions.

PRIDE AND PREJUDICE BY JANE AUSTEN

Plot: Set in rural England in the late 1700s, *Pride and Prejudice* tells the story of Elizabeth Bennet, a headstrong twenty-year-old whose repartee and independence mystify (and frustrate) nearly all those around her. Much to her chagrin, her family resides in financial turmoil and it's up to her and her sisters to settle down with wealthy gentlemen and save the family from ruin. Finding a husband is no easy feat, especially when the man you hate becomes the one you love.

Talking Points: Would you force yourself to marry someone you didn't love if the marriage could save your family from serious debt? How does the courtship of couples like Jane Bennet and Mr. Bingley compare to current dating rituals? Why do men use ridicule and condescension as a way of flirting?

Snack Suggestion: Tea sandwiches and scones.

THE YEAR OF MAGICAL THINKING BY JOAN DIDION

Plot: Didion, a successful novelist and essayist, wrote this memoir in the eighty-eight days following the loss of her husband to a massive coronary and the continued hospitalization of her daughter, first due to pneumonia and then from bleeding in her brain. Within the book's 240 pages, she questions the medical details of her husband's death, approaches her grief from both an emotional and psychological point of view, and attempts to understand why self-deception, or "magical thinking," seems to be an inexplicable part of the grieving process.

his   hers

Mr. Darcy ♂   Elizabeth Bennett ♀

"I recently purchased a storage unit to hold the books I own that no longer fit in my apartment. I have lit-themed tattoos and organize book drives. I love the characters in my favorite books more than the flesh-and-blood people around me. All of my heroes are authors, and all of my favorite quotes come from fictional characters."

Andrea Quinn
Haddonfield, NJ

Talking Points: How does Didion's understanding of "magical thinking" evolve over the course of the book? What's the difference between grief and mourning? Have you ever had to reevaluate your identity after the loss of an important relationship, whether it be with a friend, family member, or significant other?

Snack Suggestion: Chocolate-covered pretzel magic wands.

PREP BY CURTIS SITTENFELD

Plot: After Lee Fiora, a small-town girl from South Bend, Indiana, earns a scholarship to an elite boarding school near Boston, she decides to take this academic relocation as an opportunity to reinvent herself as the person she's always wanted to be—refined, fascinating, and fabulous. However, once at school, she's forced to deal with all the unwritten rules and idiosyncrasies of the modern caste system, and the person she becomes isn't necessarily the person she wants to be anymore.

Talking Points: When did you first leave your family for an extended period of time, and what did it feel like? What was your biggest insecurity during high school? Who has been your most regretful hookup?

Snack Suggestion: Cape Cod Classic Kettle-Cooked Potato Chips.

THE HANDMAID'S TALE BY MARGARET ATWOOD

Plot: *The Handmaid's Tale* takes place in the future, in the country formerly known as America, after terrorist attacks demolish the established democracy and replace it with a racist, sexist, and hierarchical dictatorship. The story follows a concubine ("handmaid") named Offred, who's forced to provide children for her owner, the Commander, and his infertile wife, Serena Joy.

Talking Points: Atwood separates the female population into various "legitimate" and "illegitimate" categories. How do those categories relate to how women are categorized today? Why are Offred's survival techniques so inward and passive? How soon can everyone get together to watch the incredibly schlocky film adaptation starring Faye Dunaway and Robert Duvall?

Snack Suggestion: Oranges.

BITTER IS THE NEW BLACK: CONFESSIONS OF A CONDESCENDING, EGOMANIACAL, SELF-CENTERED SMARTASS, OR, WHY YOU SHOULD NEVER CARRY A PRADA BAG TO THE UNEMPLOYMENT OFFICE BY JEN LANCASTER

Plot: After Jen is fired from her high-powered and high-paying job, she realizes that her high-maintenance ways—*Spa pedicures! Expensive dinners! Overpriced highlights!*—aren't going to get her very far and being jobless isn't all funemployment and games. Instead of wallowing in her bad luck and even worse attitude, Jen decides to regale the virtual world with the trials and tribulations of her everyday life, and the rest is, well, book-deal history.

Talking Points: What would you consider your most humbling life experience? What would you do if you lost your job tomorrow? Why do buses smell like body odor and corned beef sandwiches?

Snack Suggestion: Maruchan ramen.

TAP THAT SASS

For women of a certain age—namely those who were in their late teens or early twenties in the years from 1988 to 1994—being called a "Sassy girl" didn't just mean you were a precocious youth suffering from foot-in-mouth disease. It meant you listened to college radio and were interested in feminist politics. You weren't afraid to stand up for yourself or ask uncomfortable questions about sex. You were a fan of Fugazi but weren't ashamed to admit you were *kinda* obsessed with Johnny Depp. More important, it meant you were a loyal reader of *Sassy* magazine, *the* authority on girl culture.

"The test of literature is, I suppose, whether we ourselves live more intensely for the reading of it."
—author Elizabeth Drew

Steering the *Sassy* ship was founding editor Jane Pratt, a twenty-four-year-old Oberlin grad who had her chipped-polish finger on the pulse of Gen X. During her tenure at the magazine, Pratt pushed the boundaries of publishing and challenged the kind of stock editorials that *other* teen titles spoon-fed to girls. She talked to her readers like she was their therapist, big sister, and best friend. Along with the rest of her brilliant staff, Pratt focused on the idea of discovery. Whether it was a cool new band, actor, beauty product, or sexual position, *Sassy* managed to unearth the underground scene without exploiting it. No easy feat, by any means. Unfortunately, in 1994, the magazine was bought by Petersen Publishing and was eventually folded (relatively unsuccessfully) into *Teen*.

Sassy was only around for six years, but the magazine's impact can still be felt today. After being put out to pasture, Pratt went on to start her own namesake magazine, *Jane*, which carried on the irreverent *Sassy* tradition of thought-provoking editorials and down-to-earth celebrity coverage for nearly ten years. Other former *Sassy* staffers have also continued to climb mastheads at *Spin, Lucky,*

Elle Girl, and *Seventeen.* Then, in 2007, two past readers, Kara Jesella and Marisa Meltzer, hooked up and wrote the book *How* Sassy *Changed My Life: A Love Letter to the Greatest Teen Magazine of All Time,* which went on to change the lives of even more girls who weren't around to experience *Sassy* the first time around.

To this day, *Sassy* remains a secret handshake among she-geeks everywhere.

THE UNBEARABLE PRETENTIOUSNESS OF BEING

Do members of the literati really love reading male-generated postmodern* and maximalist† literature—or do they simply love *boasting* about reading male-generated postmodern and maximalist literature? Good question. To be honest, the answer probably lies somewhere in the middle. However, before you settle down with a stack of seven-hundred-page tomes, a dictionary, and a bottle of Excedrin tension headache

"Being a literary geek girl hasn't shaped my life; it *is* my life. I'm the girl who always has a notebook and pen, the girl who can argue for a half hour about a single verb and love every minute of it. I'm the girl who gets animated just talking about words, the shape of them, how they feel in my mouth. I've made all my dearest friends through writing, through reading, and through a mutual love of words. Take away literature, and I don't know who I would be."
Brittany Jo-Janssen
Holmgren
Golden, CO

---

* A post–World War II literary movement characterized by works that rely heavily on fragmented prose, paradoxes, and the reader's quest for enlightenment in an otherwise chaotic world.

† A term coined by James Wood, a critic for *The New Republic,* in reference to early twenty-first-century literature characterized by exhaustive length, manic prose, and the injection of magic into realistic settings in order to gain a deeper understanding of reality. Also known as "hysterical realism" or "magical realism."

medicine, check out this crib sheet to the top critically acclaimed authors literary geeks talk about but don't always have the time to read.

DAVID FOSTER WALLACE

Essential Work: *Infinite Jest*

What You Need to Know: Wallace is a big hit with atheists who listen to lo-fi bands like Smog and aren't intimidated by large page counts.* He is credited with being one of the forefathers of the maximalist movement. Sadly, Wallace, who suffered from severe depression for nearly twenty years, hanged himself in 2008 and was never able to finish his long-awaited third novel, *The Pale King*, although his publisher released it posthumously in 2011.

GABRIEL GARCÍA MÁRQUEZ

Essential Work(s): *One Hundred Years of Solitude and Love in the Time of Cholera*

What You Need to Know: Not only does Oprah approve of Márquez,[†] but he's also won the Nobel Prize for Literature. To achieve that kind of critical and mainstream success is almost unheard of in the literary world. Márquez is often associated with magical realism, which means he infuses elements of fantasy into everyday situations in order to reveal a deeper understanding of reality. It's kind of like tripping on 'shrooms and then doing something mundane like going to Bed Bath & Beyond—minus the potential side effects of diarrhea and needlessly buying a stainless steel chafing dish.

---

* *Infinite Jest*? More like *Infinite Page Count*. The book clocks in at 1,104 pages.
† Both *One Hundred Years of Solitude* and *Life in the Time of Cholera* were chosen as official Oprah's Book Club selections.

KURT VONNEGUT

Essential Work: *Slaughterhouse-Five*

What You Need to Know: Vonnegut purists don't often credit *Slaughterhouse-Five* as the author's best work—possibly because they feel like the story is pointless and the writing pretentious— but it's hard to argue with the antiwar novel's impact on satire and science fiction. Much like that of Mark Twain, Vonnegut's literary vision combined elements of dry wit, social criticism, and moral integrity. Plus, he's the only president of the American Humanist Association to appear in a Rodney Dangerfield movie.[*]

JONATHAN FRANZEN

Essential Work: *The Corrections*

What You Need to Know: Recognized by his signature thick-rimmed eyeglasses[†] and for being a master of postmodern suspense, Franzen's best-known work, *The Corrections*, is the literary equivalent of a Norman Rockwell painting: on the surface, it's a flawless snapshot of the carefree American family, but dig a little deeper and you'll encounter more dysfunction than on an episode of *The Steve Wilkos Show*.

> "The reader is a friend, not an adversary, not a spectator." —author Jonathan Franzen

MICHAEL CHABON

Essential Work: *The Amazing Adventures of Kavalier & Clay*

What You Need to Know: By blending together elements of Jewish mysticism, American folklore, realism, and escapism, Chabon has earned himself a reputation as one of the most important writers of the twentieth century. Nearly all of Chabon's books have been optioned by Hollywood but only one,

---

[*] Vonnegut had an inexplicable cameo in 1986's rags-to-riches comedy *Back to School*.

[†] In 2010, Franzen's specs were stolen during a London book signing and held for $100,000 ransom—until the dim-witted prankster was apprehended hours later.

Wonderboys, has successfully made it from script to screen. (Okay, that's not true. The film adaptation of *The Mysteries of Pittsburgh* was actually released in 2006 but (1) no one saw it, (2) the entire population of Pittsburgh was insulted by the production after lead actress Sienna Miller referred to their fair city as "Sh*tsburgh," and (3) did I mention that no one saw it?)

THOMAS PYNCHON

Essential Work: *Gravity's Rainbow*

What You Need to Know: Although Pynchon is a critical darling in the literary world of postmodernism—known in most realist circles as a "mathematician of prose"—he has absolutely *zero* interest in being publicly celebrated for anything. He hates talking to reporters, doesn't like getting his picture taken, and has made only two known vocal appearances, one of which was mocking his own reclusiveness on an episode of *The Simpsons*.

MAGAZINE MADNESS

The Believer

Mental Floss

The New Yorker

Slate

Utne Reader

Zoetrope

WEB BOOKMARKS

BookForum.com

GalleyCat.com

GoodReads.com

LATimesBlogs.LATimes.com/JacketCopy

McSweeneys.net

NaNoWriMo.org

PaperCuts.Blogs.NYTimes.com

Readernaut.com

TheMillions.com

GEEK-APPROVED PLAYLIST

1. "Le Pastie de la Bourgeoisie" by Belle & Sebastian (inspired by J. D. Salinger's *Catcher in the Rye*)

2. "Killing an Arab" by the Cure (inspired by Albert Camus' *The Stranger*)

3. "Don't Stand So Close to Me" by the Police (inspired by Vladimir Nabokov's *Lolita*)

4. "Lullaby" by Lagwagon (inspired by Chuck Palahniuk's *Lullaby*)

5. "Mix Tape" by Brand New (inspired by Stephen Chbosky's *The Perks of Being a Wallflower*)

6. "These Bones" by Dashboard Confessional (inspired by Stephenie Meyer's *Twilight*)

7. "Time to Dance" by Panic! At the Disco (inspired by Chuck Palahniuk's *Invisible Monsters*)

8. "Bukowski" by Modest Mouse (inspired by Charles Bukowski)

9. "The Heart Is a Lonely Hunter" by the Anniversary (inspired by Carson McCullers's *The Heart Is a Lonely Hunter*)

10. "Front Row" by Metric (inspired by Don DeLillo's *Great Jones Street*)

11. "Cherry Lips" by Garbage (inspired by J. T. Leroy's *Sarah*)

12. "The End" by My Chemical Romance (inspired by William Faulkner's *A Rose for Emily*)

13. "Resistance" by Muse (inspired by George Orwell's *1984*)

Three

FILM GEEK

Do you know the difference between Woody Allen and a woodentop?[*] **If you're unsure, then put down the remote and test your film geek girl skills!**

1. Which is the only Wes Anderson–directed movie Bill Murray has *not* appeared in?

    A. *The Life Aquatic with Steve Zissou.*

    B. *Rushmore.*

    C. *Bottle Rocket.*

---

[*] According to TheFilmGeek.org, a woodentop is "a performer who, for some reason, commands large paychecks even though there seems to be absolutely no sign of any acting ability whatsoever." Notable examples include Vin Diesel, Brendan Fraser, and Jennifer Lopez.

2. For years, many cinephiles have argued over the symbolism of the closing sequence in the 1999 movie *Magnolia*, which features hundreds, if not thousands, of frogs raining down from the sky. The real reason why director Paul Thomas Anderson chose to end the film with a frog storm is because . . .

> A. Beats me. Why don't you ask Robert Altman? After all, director Paul Thomas Anderson totally ripped off the movie from Altman's 1993 multicharacter drama *Short Cuts*.
>
> B. It was way cheaper for director Paul Thomas Anderson to rent a bunch of rubber frogs instead of trying to invest in larger, more traditional animals that rain from the sky, like cats and/or dogs.
>
> C. Inspired by the supernatural-phenomena writing of Charles Fort, director Paul Thomas Anderson wanted to draw attention to the notion that the health of society can be judged by the health of its frogs. The frogs tell us who we are, as people; thus, when we pollute our frogs, we're actually polluting ourselves. The whole occurrence is most likely a reference to Exodus 8:2, which states, "And if thou refuse to let them go, behold, I will smite all thy borders with frogs."

3. You just found out that *Looking for Mr. Goodbar*—the 1977 sexual thriller starring Richard Gere (before he hit his silver-fox stride) and Diane Keaton, who went on that same year to win an Oscar for her role in *Annie Hall*—is finally going to be released on DVD.* In anticipation of its release, you . . .

> A. Immediately add the title to your Netflix queue because you can't pass up the chance to watch Richard Gere do push-ups in his jock strap—and in HD, no less.

---

* Sadly, it's not, but a girl can dream, can't she?

B. Are inexplicably salivating over the thought of chocolate and peanuts, so you put on your shoes and decide to hoof it to the nearest 7-Eleven for a snack.

C. Wave your fist and curse the heavens because everyone knows that a movie doesn't have a *proper* DVD release unless it's through the Criterion Collection. *Drat!*

4. Alfred Hitchcock is to film noir as Jean-Luc Godard is to:

A. Neo noir.

B. General Foods International Coffee.

C. Cinéma vérité.

5. Before penning the script for *Juno*, Diablo Cody was your average sassypants writer living and blogging in Minneapolis. She was officially discovered when a Hollywood heavyweight started reading about her adventures and loved her voice, tone, and snark. Her blog, called *The Pussy Ranch*, was about . . .

A. Cody's obsession with cats in cowboy hats.

B. Cody's new line of vegan, low-calorie salad dressing.

C. Cody's exploits in the local Minneapolis sex industry, which was later hilariously documented in her book *Candy Girl: A Year in the Life of an Unlikely Stripper*.

ANSWER KEY

Mostly As: I really appreciate your fervor and excitement, but just because you've watched *Fargo* doesn't mean you're a film geek girl . . . *yet.* Read on and get yah Netflix queue ready.

Mostly Bs: Sorry, Megan Fox, but having a gigantic Marilyn Monroe tattoo on your inner forearm doesn't make you look like a fan of classic cinema. In fact, it just makes you look like a poseur. Oh, did you hear that? I think Brian Austen Green's calling you from the other room. *Beverly Hills, 90210* is playing on the Soap Network right now, and he wants you to keep him company while he weeps uncontrollably over his lost youth.

Mostly Cs: Mazel tov, maven! You're a bona fide film geek girl!

CHARACTER SKETCH

From an early age, film geek girls realize that they're not like everyone else. When they're li'l tots and their parents sit them down in front of kid-tested, mom-approved films like *Beauty and the Beast* and *Harriet the Spy*, they aren't just cooing at the pretty princesses and laughing at the calculated pratfalls; instead, they're contemplating the advances in Disney animation since *The Little Mermaid* and wondering who adapted the *Harriet* screenplay from the original Louise Fitzhugh–penned book. Well, maybe that's a slight stretch, but there's no doubt that a passion for cinema develops early in film geek girls.

It's this authentic affection and obsession for film that separates the moviegoer from the movie-lover. See, moviegoers enjoy watching flicks with their friends, stuffing their faces with handfuls of oversalted popcorn while laughing and crying at all the appropriate parts. But when they walk out of the theater, the movie is all but a distant memory. For movie-lovers, the experience couldn't be more different. Sure, they recognize that going to the movie theater is often social in nature, but the real fun happens after the film ends. That's when they can dissect and discuss what they've seen, thus infinitely extending the movie-watching experience and allowing them to escape further and deeper in the cinematic abyss.

For this reason, film geek girls have a tendency to be hyper self-aware, which can often lead to thoughts like, "Am *I* normal?" "What *is* normal?" "Do I even want to *be* normal?" Not only can such internal dialogues drive a girl into existential overload, but long-winded discussions with oneself can be utterly exhausting. It's no wonder that, at the end of the day, these she-geeks prefer to silence the voices by escaping into someone else's world for eighty-eight minutes. Sometimes our armchair artists feel safer as a visitor in a foreign cinematic land than as an active participant in their own reality. Hell, real life can be *scary*. When you

"You know how you're always trying to get things to come out perfect in art, because it's real difficult in life." —Alvy Singer, in *Annie Hall*

"Movies allow me to escape from reality for an hour and a half. I love diving deep into the stories and looking at the details that nobody takes the time to notice. To me, a good film is one that has me still thinking about it long after the credits have rolled."
Andrea Eastman
Columbus, OH

have something to say, you can't always find the perfect words, and when you want to take a risk, you can't always work up the nerve. That's why watching films can also be cathartic for our she-geeks; they can watch their onscreen counterparts do and say all the things they wish they could in real life.

Much like playing peewee hockey and collecting baseball cards, obsessing over films is oftentimes considered to be a male pastime. That doesn't mean there isn't room for geek girls in this boys' club. In fact, these intellectual tomboys are usually welcomed with open arms—and lots of conspiracy theories about why *Margaret*, the follow-up film from director Keith Lonergan (*You Can Count on Me*), was filmed in 2005 and *still* hasn't been released. Unfortunately, be forewarned if you're crushing on your film buddy: many times such gals are prematurely banished to the "friend zone." In a film geek guy's head, there are two types of girls: the ones you want to converse with *after* the movie and the ones you want to make out with *during* the movie. I mean, who would *you** want to suck face with—Diablo Cody or Amanda Seyfried?

That doesn't mean all is lost, my fellow film enthusiasts. It just means you might have to be patient before your milkshake brings all the boys to the yard. Fortunately for us, like a fine bottle of Francis Ford Coppola wine, film geek girls get better with age.

"I'm just trying to tell a story, and my hope is that though what I went through is very personal, it's also very universal." —*Tiny Furniture* director Lena Dunham

GEEK MYTHOLOGY

Films have made such a *huge* impression on the modern geek girl, it's easy to forget that filmmaking has only been around for the past hundred years or so. In the creation of any art form,

---

* By "you," I mean "pimply-faced teenage boy who plays lacrosse, lives to make his own Frappuccino at Starbucks, and spends his Friday nights flashing people on Chatroulette."

there are going to be a ton of firsts, especially when said art form evolves alongside the women's suffrage and equal rights movements of the twentieth century. For example, **Alice Guy-Blaché** was the first woman to open and operate her own production studio, many consider **Frances Marion** to be the first successful female screenwriter, and **Lois Weber** is credited with being the first woman to direct a full-length feature film—and that was all *before* 1930.

As the women's liberation movement progressed over the years, so did the female contingent in Hollywood. Although **Ruth Gordon** had been writing screenplays since the 1940s, it wasn't until she appeared in *Rosemary's Baby* (as Minnie Castevet, the incredibly nosy—and creepy—next-door neighbor) and *Harold & Maude* (as Maude, who was a proponent of May-December romances *waaay* before Ashton and Demi made it trendy) that her unabashed eccentricity resonated with audiences. Then there's the incomparable renaissance geek **Barbra Streisand**, who never met a musical she didn't like. Sure, her performances in *Funny Girl* and *The Way We Were* earned her critical praise, but audiences appreciated her nonconformity more than anything else. Though she was pelted with insults for being "unusual" and "unattractive," no one ever dared to call her untalented. Despite the haters, Babs never felt the need to fit the Hollywood mold. Instead, she made the Hollywood mold fit *her*.

In 1977, **Diane Keaton** starred as the title character in Woody Allen's *Annie Hall* and geek girls everywhere rejoiced because they had finally found their cinematic role model. The movie was about a New York City couple who engages in an on again, off again romance that seems more like an emotional car crash than a tender love affair. Self-deprecating, neurotic, and utterly tomboyish, Annie (and Keaton, more importantly) was the type of gal other flawed females could identify with—and look up to. After all, cinematic escapism isn't all about fleeing reality; some-

"People ask me, 'How can *you* be making movies about men? When you're, you know, a woman.' I recently read that [director] Mike Leigh gets the same kind of question about making movies about women. He gets really pissy and says, 'I make movies about people.' And that's a perfect answer. There's this theory of multiple intelligences, and I may not be smart on many of the others, but my emotional intelligence is pretty high up there. I've always been a close observer of human beings and what makes them tick, regardless of gender." —*Humpday* director Lynn Shelton

times it's just about watching a better, different, or perfectly lit version of yourself.

By the time the 1980s came around audiences seemed ready for a reality check, and a pack of bratty actresses were about to give it to 'em. It's impossible to talk about movies of the decade without discussing **Molly Ringwald**, the face that launched a hundred John Hughes films.* Over a matter of three years and just as many flicks (1984's *Sixteen Candles*, 1985's *The Breakfast Club*, and 1986's *Pretty in Pink*), Ringwald managed to create an authentic blueprint for teen-girl angst, especially for girls who stood on the outside of popularity looking in. Even when she was playing Claire Standish, Miss Popular in *The Breakfast Club*, she still acted as though it could all fall apart at any second—no thanks to the outsider taunts of Allison Reynolds (played by **Ally Sheedy**), a social pariah with a compulsion for lying and an incredibly dry scalp. And though she never appeared in any of Hughes's movies, **Lili Taylor** earned her share of geek merit badges for her roles in films like *Mystic Pizza*, *Say Anything*, and *Dogfight*.

---

* Technically, it was only three, but you get the point.

Before she went on a shoplifting spree at Barneys and started costarring in really bad Adam Sandler movies, **Winona Ryder** was poised to be crowned geek-girl royalty: she starred in cult classics like *Heathers*, *Reality Bites*, and *Girl, Interrupted*, dated geektastic rockers, and proved you didn't have to be Kate Moss to pull off the whole "waif" look. Unfortunately, Ryder's star waned by the turn of the century, and she struggled to regain her geek cred until appearing in 2010's *Black Swan*. (When journalists asked her about the unfortunate events leading up to her acting hiatus, she blamed her breakdown on a battle with anxiety and depression. I blame it on a delayed reaction to the horrible dialogue in *Autumn in New York*. That drivel would send anyone into an uncontrollable shame spiral.)

Regardless, a new wave of film geek girls is ready for their close-up, Mr. DeMille. **Kat Dennings** keeps it real onscreen in such geek-approved films as *Daydream Nation* and *Nick and Norah's Infinite Playlist* and online, by personally running her own namesake website and Twitter. Another honest-to-blog film geek girl is **Olivia Thirlby**, who earned a lot of unique visitors by playing sidekick to fellow she-geek **Ellen Page** in *Juno* and love interest of Josh Peck in the hip-hop indie drama *The Wackness*. It's hard to find a girl with a geekier film résumé than **Anna Kendrick**, who can brag about starring opposite Robert Pattinson (in *The Twilight Saga*), Michael Cera (in *Scott Pilgrim vs. the World*), and George Clooney (in *Up in the Air*). Finally, there's **Emma Stone**, whose incognito geekiness in *Superbad*, outstanding zom-com timing in *Zombieland*, and comic book cred in the *Spider-Man* reboot almost make up for the fact that she was in *Ghost of Girlfriends Past*. Almost.

"As a result of watching so many movies, I've trained myself to imagine the painful or boring portions of my life as montages, which makes them a lot more bearable."

Katie Wright
New York, NY

FILM GEEK GODDESSES

It's an exciting time to be an out-and-proud film geek girl. Not only are the latest crop of It-girl actresses more authentic than ever but sister-friend directors like Kathryn Bigelow, who was the first woman to win the Academy Award for Best Director for *The Hurt Locker*, and Catherine Hardwicke,* who was responsible for bringing the first installment of *Twilight* to cinematic life, are inspiring a new generation of ladies to man the lenses. You'll notice that the majority of the following goddesses tend to dominate independent films, although all have dipped their toes in the occasional superhero summer blockbuster or Nancy Meyers–penned romantic comedy. That's not to say that film geek girls are only found on indie sets; however, lower-budget movies tend to give our goddesses the perfect platform to explore their quirk and charm. Now, before we lose this sunlight, can someone tell my first A.D. to grab the clap slate so I can yell "Action" on my picks for today's top film geek goddesses? Ugh . . . *amateurs.*

SOFIA COPPOLA, Voted High Style, High Art Indie Film Geek

Audiences didn't expect much from Sofia Coppola after she was pelted with metaphorical tomatoes for her dreadful acting performance as Mary Corleone in *The Godfather: Part III*, but the daughter of famed director Francis Ford Coppola wasn't ready to say arrivederci to the cinema just yet. Instead, she took her fated place behind the lens and helmed some of the best art house films of the last ten years—*The Virgin Suicides*, *Lost in Translation*, and *Marie Antoinette*. Dripping with untraditional beauty and unrivaled coolness, Sofia has inspired a fleet of burgeoning geek

---

* There are also a lot of inspiring film geek gals who aren't named Kathryn/Catherine, so don't get discouraged if you didn't happen to be born with that moniker.

girls who believe that independent film can have both style and substance. If I was forced to describe her in seven words, I'd say she's "revolutionary, petulant, reactionary, ebullient, fragrant, cold [and] cool"—at least that's what it says on the label of her namesake sparkling wine for Coppola Winery.

ZOOEY DESCHANEL, Voted Doe-Eyed, Sweet-as-Pie, Perfectly Imperfect Film Geek

Named after the lead character in J. D. Salinger's novella *Franny and Zooey*, Zooey Deschanel seemed destined for geek greatness since birth. However, it was roles in movies like *Almost Famous*, *All the Real Girls*, and, more recently, *(500) Days of Summer* that solidified her film goddess status. She's got indie cred coming out the wazoo, yet she maintains a level of humility and niceness that seems almost inhuman. There's an innocence and eagerness that she brings to every performance—whether it's onscreen or onstage, singing with her folk-rock band She & Him—and you can't help but be smitten by the fact that she seems completely unaware of her loveliness. Ignorance might not be bliss, but it's definitely something that many film geek girls suffer from. Maybe Zooey and her husband, Death Cab for Cutie frontman and prince of the music geek guy kingdom Ben Gibbard, will eventually write a song about it. I bet it would be a toe-tapper.

> "There's nothing else like the experience of sitting in a movie theater, watching something that moves you, thrills you, makes you cry, makes you laugh, makes you want to be more than you are."
>
> *Jaime Smith*
> *Warwick, RI*

MAGGIE GYLLENHAAL, Voted Indie-lectual Feminist Film Geek

Aside from a single forgettable role—as Shannyn Sossamon's bestie—in an even more forgettable film—the screwball-slash-hornball rom-com *40 Days and 40 Nights*—Maggie Gyllenhaal is one smart cookie who's not afraid to boast her intelligence on- and offscreen. This cerebral sex symbol tends to submerge herself in pieces of cinema that stimulate (e.g., *Adaptation*), challenge (e.g., *Stranger Than Fiction*), and engage (e.g., *Crazy Heart*) the senses. Oh, and don't get me started on her role as Lee, the lovesick, OCD-stricken receptionist in *Secretary*. If I may be so bold, that film is the equivalent of geek porn—or at least the kind of porn that geeks will admit to watching—and I can only imagine the sales of bondage bars skyrocketed that year. Not like I know or anything . . . *(Awkward silence.)*

SARAH POLLEY, Voted Anti-Hollywood Hollywood Ingénue Film Geek

When Sarah Polley first debuted to American audiences, it was in the rave-caper comedy *Go*, where she played Ronna, a down-and-out grocery store checkout girl who attempts to sell drugs in order to pay for her overdue rent. Besides Katie Holmes appearing almost human in the movie, Polley was definitely the most memorable thing about the film. However, instead of heading down a yellow-brick road paved with romantic comedies and summer blockbusters, Polley dismissed the idea of Hollywood fame in favor of more low-budget, obscure, independent fare like the tragic dramas *My Life Without Me* and *Away from Her*, which she also directed. What I love about Polley is her refusal to become just another faceless starlet—even when it seems to be the easiest and most profitable thing to do. Lesson learned: a girl's geekiness might cost her popularity points, but there's more to life than having a bunch of Facebook friends and nameless Twitter followers—especially when your misfit chutzpah might earn you an Oscar nomination later in life.

NATALIE PORTMAN, Voted Beauty and Brains Film Geek

Fangirls first initiated Natalie Portman into the geek sisterhood when she signed on for the role of Padmé Amidala in the *Star Wars* prequels, but she earned her film-geek icon status when she played the more relatable—yet completely eccentric—character of Samantha in *Garden State*. As if that weren't enough, Natalie is also an unapologetic academic. Sure, we'll never know if she got into Harvard because she had jaw-dropping SAT scores or because she may've gotten a letter of recommendation from Michael Mann, but at least we know she's got something to fall back on in case this whole acting thing doesn't work out.* Plus, I gotta give big ups to Natalie for not taking herself *too* seriously. (I mean, did you hear her über-geeky giggle when she accepted the Best Actress Golden Globe for her role in *Black Swan*? That was a meme waiting to happen.)† After all, when you're constantly searching for cinematic enlightenment, it's easy to forget how important it is to stop every once in a while to smell the roses—or drop a couple of f-bombs while spitting some highly offensive gangsta-rap rhymes on *Saturday Night Live*.

"There's a difference between being in a bra and underpants as an object on a men's-magazine cover and playing yourself—a woman with desires and needs who loves and laughs with her friends—in a bra and underpants. Most movies are made by men, it's totally natural that they're going to present their worldview, so we're trying to find more woman who are writers and directors who are expressing their worldview."
—actress Natalie Portman, on the goal of her production company Handsomecharlie Films

---

* She graduated from Harvard in June 2003 with a bachelor's degree in psychology. Dr. Portman, I presume?

† And leave it to the guys at CollegeHumor.com to turn the twelve-second chortle into a three-minute supercut.

HALL OF FAME: PARKER POSEY

When I first saw Parker Posey pop up onscreen as Mary, the downtown NYC club kid who turns into an uptown librarian in the cult classic film *Party Girl*, she had me at "He-he-hello!" From there, my love only grew after I saw her take geektastic turn after geektastic turn in movies like *Basquiat, House of Yes,* and *Clockwatchers*. Rail thin with a razor-sharp tongue, Posey is often cast as a no-nonsense woman-on-the-verge. "Of what?" you ask. Well, it all depends on the project, but that's secondary to the fact that when you're a geek girl, it feels like you're always on the verge of something, whether it's a breakdown or a breakthrough. Truth be told, sometimes our passions can be too much to take—like that line from *My So-Called Life* where Angela Chase says, "You're so beautiful; it hurts to look at you."

In terms of Parker Posey, her onscreen neuroses are admirable. Yes, some have been far-fetched, like sporting adult braces and an unhealthy obsession with her Weimaraner in *Best in Show*; but some hit almost too close to home. (Two words: *Broken English*.) Back in 1997, Parker was hailed by *Time* as "Queen of the Indies." I'd like to suggest that her reign should also extend over film geek girls, as well. That's one kind of elective monarchy I could totally get behind!

FRENEMIES

Just because you've been a member of Netflix since 2003, own the Criterion Collection edition of *Chasing Amy,* and know how to properly pronounce Jean-Luc Godard's last name,* you're not automatically a film geek girl. As we've discussed, there's a fine line between being a movie watcher and a movie lover. If you're

> "I couldn't get an interview even though my last movie made $400 million. I was told it had to be directed by a man. Am I crazy? [*The Fighter*] is about action, it's about boxing, so a man has to direct it. . . . But they let a man direct *Sex and the City,* or any girly movie you've ever heard of." —*Twilight* director Catherine Hardwicke on being turned down to direct *The Fighter*

---

* Which would be "Goh" + "dahr," for those who aren't hooked on phonics.

truly in the second category, you'll have to look out for these fren-emies. They might talk the talk, but they definitely *don't* walk the walk—or watch the bonus disc with audio commentary.

- ✔ The poor saps who wait in line for hours at "lifestyle centers"* like The Grove in Los Angeles, California, or Easton Town Center in Columbus, Ohio, to see films like *Supercross*, *Dear John*, or anything starring Channing Tatum.
- ✔ People who download movies through BitTorrent sites like the Pirate Bay, LimeWire, and Gnutella.
- ✔ Anyone who refuses to discuss the meaning of *Mulholland Drive* because "overanalysis diminishes the film's effectiveness and strips the viewer of the true cerebral experience." *Douche.*
- ✔ Cheeseballs who still quote *Napoleon Dynamite*, *Borat*, or *Austin Powers* on a regular basis.
- ✔ *Transformers* director Michael Bay.
- ✔ Buyers who don't understand why Blu-ray is so much better than normal DVDs.†
- ✔ Viewers who think the films shown on the Lifetime Movie Network are examples of highbrow cinema.‡
- ✔ Poseurs who admit to being "really into film" after seeing *one* Wes Anderson movie.

"I think bad laughers are a movie lover's worst enemy. There's always that one super-loud person who laughs at *everything*. You know, the person who ends up sitting right behind you in the theater, laughs at things that aren't even funny, and then laughs extra hard at the things that are, so much so that you can't hear the next few lines. It's never a charming laugh, either. It's always the most annoying laugh you've ever heard."

Christina Johns
Glendale, CA

---

* Outdoor malls that may—or may not—include housing. Ever wanted to live at the mall? Thanks to these "lifestyle centers," now you can! Goodie. *(Puts hand up to head, makes gun motion, and pulls imaginary thumb trigger.)*

† All I know is it has to do with lasers that enable the Blu-ray discs to hold tons more information than your normal DVD disc. Lasers are cool.

‡ Although *She's Too Young*—a harrowing tale of promiscuous teens who partake in risky sexual behavior that leads to a chlamydia outbreak at a local high school—starring Academy Award–winning actress Marcia Gay Harden *is* pretty riveting.

THE *REAL* ROM-COM QUEENS

Sure, millions of moviegoers run to the theater in droves to see the latest piece of romantic-comedy garbage starring actresses like Kate Hudson, Jennifer Aniston, or Katherine Heigl. But, in real life, the girl rarely ends up with Matthew McConaughey at the end of a tumultuous relationship. Instead, she's usually left with a load of heartbreak, a mountain of empty Ben & Jerry's containers and a piss-poor credit score—that is, if her ex was a real gem and emptied out her bank account before he left town. Maybe one of the best things about the movies is that, unlike those of us watching from the couch, these onscreen antiheroines manage to open a can of intellectual whoop-ass while maintaining a certain geek *je ne sais quoi*. Suck it, Julia Roberts!

ENID COLESLAW (Played by Thora Birch in *Ghost World*)

Based on the satirical graphic novel by Daniel Clowes, *Ghost World* is the story of two disaffected teens—sharp-tongued Enid Coleslaw and her slightly less cynical BFF Rebecca Doppelmeyer—as they deal with the ups and (mostly) downs of life after high school. Enid is as abrasive as they come and offends nearly everyone she meets with her unfiltered dialogue and brutally honest opinions—everyone, that is, except forty-something Seymour, a record-collecting introvert who has zero friends and even less self-esteem. Enid might not look like your average silver-screen siren,[*] but it's her unabashed individuality and rough-around-the-edges demeanor that ultimately makes her so inexplicably attractive. Sure, she uses her geekiness as a shield to keep people away, but even the most powerful onscreen geek heroines have a romantic Achilles heel; hers just happens to be meek, passive-aggressive, and socially inept losers.[†]

DIANE COURT (Played by Ione Skye in *Say Anything*)

I don't care how black or cold your heart is, *every* geek girl dreams of a guy like Lloyd Dobler standing outside her bedroom window, holding a boom box, blasting Peter Gabriel's "In Your Eyes." Yes, ma'am. Diane Court was one lucky lady—one smart, quirky, average-looking, lucky lady. She single-handedly gives hope to all the academic geek girls out there, proving that we bookworms might be better romantically served by spending our time getting all As instead of saving up for a pair of double Ds. Diane Court is cinematic proof that nice girls can finish first and—(*air quotes*)—snag the guy, too.

CLEMENTINE KRUCZYNSKI (Played by Kate Winslet in *Eternal Sunshine of the Spotless Mind*)

Clementine Kruczynski—with her Kool-Aid–colored hair and flair for the dramatic—gets to live out a fantasy life

---

[*] After all, how many female lead characters don a vinyl Catwoman mask without being . . . well . . . Catwoman?

[†] I can relate, Sister Sledge.

that most of us mere mortals only dream about. Not only does she have the presence of her bitter ex-boyfriend Joel Barrish (Jim Carrey) surgically erased from her memory, but no sooner does she wipe the romantic slate clean than she's pursued by an unusually creepy Elijah Wood. (Sure, he's siphoning all her ex's intimate memories in order to win her heart, but it's *effing* Frodo. Who *wouldn't* hit that? I mean, really.) Unfortunately, Clementine capriciously deleted her failed relationship, and the spiteful act nearly destroys Joel; however, it's the same impulsivity that initially attracts, intrigues, and captivates him. Clementine is a pistol, a firecracker, and, more important, an acquired taste. But, then again, she never tries to be Miss Popular. In fact, she sort of thrives on being a polemicist,* which is a very film geek girl thing to do. Why fit in when you can stand out?

> "Look, man, I'm telling you right off the bat I'm high maintenance. So I'm not gonna tiptoe around your marriage or whatever it is ya got goin' on there. If you wanna be with me, you're with me." —Clementine Kruczynski to Joel Barrish in *Eternal Sunshine of the Spotless Mind*

JUNO MACGUFF (Played by Ellen Page in *Juno*)

Ah, yes. Who can resist the female smart-ass with the heart of gold? Not me and definitely not the zillions of home-skillets who flocked to the theater to see *Juno*, the coming-of-age dramedy about a knocked-up teen and her quest to find suitable parents for the unplanned bun in her oven. Many detractors criticized Juno's signature slang because unless your name is Diablo Cody and you reside in the United States of Tara, no one *actually* talks like that 24/7. But wouldn't it be a better world if they did? Even if you duct-taped her mouth shut, Juno would still represent the kind of onscreen heroine that we don't often see in big-budget Hollywood flicks: she's kind of a tomboy, looks like

---

* This is basically a fancy word for "someone who likes to argue for the sake of arguing."

she hasn't washed her hair in days, thinks Sonic Youth tends to sound like "just noise," drinks Blue Shock Slurpees by the gallon, and likes a dude who wears thin-wale corduroys and has zero vocal inflection. In other words, she's just like us! (Insert *Us Weekly*–style enthusiasm here: e.g., *She pumps her own gas! She gets parking tickets! She picks a wedgie!*) Plus, I've watched enough episodes of *16 and Pregnant* to know that it's always the incognito wallflowers who end up getting the most action—and, apparently, don't know how to use a condom. *(Rimshot.)*

> "Love is mysterious and rad, like Steve Perry from Journey." —screenwriter Diablo Cody

MARGOT TENENBAUM (Played by Gwyneth Paltrow in *The Royal Tenenbaums)*

There's something inexplicably alluring about Margot Tenenbaum, whether it be her stoic demeanor, her missing finger, or her overapplication of dark eye makeup. Tall and statuesque, she seems to be followed by a cloud of mystery everywhere she goes—though her hair may not be full of

secrets (in actuality, it only holds two small barrettes). The less you know, the more you want to find out. It's no wonder that every man that meets her seems to fall at her feet—even her own brother . . . But she's adopted and, more important, he's Luke Wilson, so it's totally copacetic. Whereas many film geek girls get unfairly categorized as being somewhat nerdy and neurotic, Margot represents the cool, calm, cerebral alternative. Thankfully, even if you *are* nerdy and neurotic, you can dress up as Margot Tenenbaum

for Halloween, experience how the other half lives, and, fingers crossed, silence the voices in your head for at least one night.

GEEK LOVE

"My Netflix instant queue is my boyfriend when my husband is out of town."

Jessica Johnson
Brooklyn, NY

When most girls dream of their Prince Charming, they usually envision someone who's tall, dark, and handsome. But that type of guy would be an absolute nightmare for any film geek female. Sure, she wouldn't kick a babeasaurus like Chris Evans out of bed, but there's a *slim* chance he'd be down to catch a matinee of *Stella Dallas* at the Silent Movie Theatre the following day. Then again, she doesn't want someone *so* cinematically belligerent that any ol' film discussion turns into *The War of the Roses.** There's got to be a middle ground.

On that middle ground stands a potential paramour who would rather spend a beautiful summer day holed up watching a Coen brothers movie marathon than playing volleyball on the beach; he doesn't understand why people find Holly Madison attractive; and he certainly doesn't see the need to spend more than $100 on jeans when you can buy a perfectly good pair of Levi's at Goodwill for less than a quarter of that price. However, film geek girls should be cautious of those mad-genius James Cameron types. They tend to recycle relationships faster than *American Pie*'s last straight-to-DVD release—and that's one kind of "hurt locker" that no girl wants to sit through.

---

* Not like the dude would've seen *The War of the Roses.* After all, James L. Brooks is *waaay* too Hollywood for his refined taste.

THE PERFECT MATCH FOR A FILM GEEK GIRL . . .

❑ Looks like Ben Foster, Joseph Gordon-Levitt, or Mark Ruffalo.

❑ Prays at the altar of Netflix instant streaming.

❑ Understands the difference between a jump cut,* a match cut,† and a smash cut.‡

❑ Refuses to walk out on a film, no matter how horrible it is.

❑ Is more interested in making Oscar predictions than participating in his friends' March Madness college basketball bracket.

❑ Has an accent.

❑ Is not interested in sports or movies about sports.

❑ Knows how to hook up your laptop to your TV—and possibly carries around a spare VGA cable in his back pocket.

---

* A jarring editing technique where the middle part of a continuous-action scene is cut out.

† An editing technique where two unrelated shots are joined together to make them appear to be related.

‡ An editing technique where the transition between shots is abrupt and meant to startle the viewer.

WEB BOOKMARKS

Deadline.com

Defamer.com

EW.com

HollywoodReporter.com

IMDb.com

JoBlo.com

Movieline.com

MUBI.com

Netflix.com

RottenTomatoes.com

TheGirlsOnFilm.com

MUST-SEE MOVIES

(500) Days of Summer

Annie Hall

Best in Show

Breakfast at Tiffany's

Casablanca

Citizen Kane

Election

Garden State

Girl, Interrupted

The Godfather trilogy

Gone with the Wind

The Graduate

Heathers

Juno

Lost in Translation

Network

Nick & Norah's Infinite Playlist

Persepolis
Pretty in Pink
Reality Bites
Rushmore
Secretary
Slums of Beverly Hills
The Squid and the Whale
Sunset Boulevard
Welcome to the Dollhouse

PEEPS TO FOLLOW ON TWITTER

Kat Dennings (@officialkat)
Zooey Deschanel (@therealzooeyd)
Lena Dunham (@lenadunham)
Roger Ebert (@ebertchicago)
Carrie Fisher (@carrieffisher)
David Lynch (@david_lynch)
Kevin Smith (@thatkevinsmith)
Peter Travers (@petertravers)
Edgar Wright (@edgarwright)

Four
MUSIC GEEK

Not sure if you've got your doctorate in the area of geek girl musicology? Take off your headphones, pick up a pencil, and test your she-geek skills!

1. Back in the 1990s when MTV used to show actual videos, the channel had a show called *120 Minutes*, which featured clips from all sorts of alt-rock, grunge, and punk bands. Which former *Rolling Stone* staffer was the first female host for the ninety-minute video block?*

    A. Kennedy.

    B. Daisy Fuentes.

    C. Jancee Dunn.

---

* Oddly enough, a show called *120 Minutes* was only ninety minutes long. Yeah, I don't get it, either.

2. The outside of Solutions Audio-Video Repair in Silver Lake, California, has become the unofficial memorial site for deceased singer-songwriter Elliott Smith, who was poignantly photographed in front of the store's red- and navy-splashed wall for the cover of his 2000 album *Figure 8*. Which self-taught female photographer took the famous shot?

> A. Former *Rolling Stone* staff photographer Annie Leibovitz.
>
> B. Tyra Banks.
>
> C. Autumn de Wilde. In fact, de Wilde even released a book of photography from the time she spent with Smith while he was recording *Figure 8*. The volume also features handwritten lyrics, interviews with Smith's friends and collaborators, and a CD of previously unreleased live material.

3. In 1989, Sonic Youth bassist Kim Gordon launched her own girls' streetwear line. What was it called?

> A. Goo.
>
> B. Sonic Threads.
>
> C. X-Girl. The brand was initially launched to be like the sister line to X-Large, the hip-hop-inspired clothing label heavily influenced by Beastie Boys' Mike D. At the time, a then-unknown underground It-girl named Chloë Sevigny was the fit model, and X-Girl apparel appeared in videos by the Breeders ("Cannonball"), the Lemonheads ("Big Gay Heart"), and, of course, Sonic Youth ("Sugar Kane").

4. Which of the following is *not* the name of a band featuring the vocal stylings of tough-girl singer Alison Mosshart?

> A. Discount.
>
> B. The Kills.
>
> C. I Can't See Through My Bangs.

5. The post-punk band Jawbreaker reached cult status in the mid-1990s after releasing an album that, at the time, was considered a huge mainstream and critical flop. However, in the years since, this particular album has gone on to inspire an entire generation of indie rockers. Name this album.

    A. *Perfecting Loneliness.**

    B. "Hold the phone . . . I thought *Jawbreaker* was a black comedy starring Rose McGowan. Are you trying to tell me that Rose McGowan was in a band, too? This is information overload. I think I'm going to need to sit down."

    C. *Dear You.*

ANSWER KEY

Mostly As: If I were to issue you a music geek girl report card, it'd probably say something like: "Enthusiastic about participating, but needs to develop background knowledge and comprehension of subject material." In other words, you'll need to enroll in summer school before I can let you graduate with the rest of your class.

Mostly Bs: Sorry, Taylor Swift, but just because you follow New Found Glory's Chad Gilbert and Paramore's Hayley Williams on Twitter *doesn't* give you street cred. Don't you have a duet with Stevie Nicks to butcher or something?

Mostly Cs: Kudos, kiddo! You're a bona fide music geek girl!

---

* This is actually the name of the last album from Jets to Brazil, the band frontman Blake Schwartzenbach formed after Jawbreaker.

CHARACTER SKETCH

Ask these girls when they first fell in love with music, and our she-geeks can usually pinpoint the year, month, day, time, and song that's associated with their discovery. For many, such audio epiphanies take place during childhood. For example, perhaps they're riding in the backseat of their family car while their dad drives and plays air drums along with Booker T & the MGs' "Green Onions" blasting from the oldies radio station. On the other hand, it might take years for you to hear the song that changes your life—not until that cool senior on the field hockey team makes you a mixtape and it starts with the Pixies' "Wave of Mutilation," which inexplicably ignites butterflies in your stomach.

After all, music isn't a part of the music geek's life—it *is* their life. And, for that very reason, every event in a music geek girl's existence—from the mundane to the extraordinary—comes with a soundtrack. Some people keep photo albums or scrapbooks to remember the good times; these she-geeks keep mixtapes and playlists. This way, a memory is just a song away—and that's the perfect distance, if you ask me.

Girls in this geeky subset can often be further distinguished by musical genre. For example, indie-rock aficionados often sport Built by Wendy–style frocks or equally minimalist ensembles, little to no makeup, and Nico-inspired hairstyles; electroclash lovers prefer neon colors, metallic appliqués, and asymmetrical haircuts; emo loyalists usually fill their closets with threads from Hot Topic and circle their eyes with massive amounts of black liner; and punk-rock devotees find solace in jean jackets with studded back patches and safety pins sticking out from every orifice.[*]

---

[*] Caveat emptor: The above descriptions are merely guidelines to understanding the music geek girl's aesthetic. Just like the exclusive vinyl offered through Third Man Records' Vault membership, female audiophiles come in all different shapes, sizes, and colors.

Music geek girls are escapists by nature. Whether they're dissatisfied with the status quo of family, friends, school, or work, these gals seek peace in a playlist and solace in a soundtrack. When the going gets tough, the tough put on headphones, cue up their stereo, and dive into the musical abyss. For this reason, these she-geeks are assumed to be loners and introverts; however, that's not always the case. Music geek girls often have tons of friends—many of whom are equally obsessed with the same bands and albums—but they might tune out when they feel emotionally overwhelmed. Lips are sealed, doors are shut, and instead of confiding in those around them, our geekettes turn off the outside world, turn up the volume on their favorite song, and allow the voice on the other side of the speaker to be their best friend, confidant, and therapist.

GEEK MYTHOLOGY

By my calculations, modern music geekdom dates back to the early 1960s with the unprecedented songwriting skills and untraditional beauty of **Carole King**, who not only penned some of the decade's most memorable Motown hits—like "Will You Still Love Me Tomorrow?" and "One Fine Day"—but also went on to have a supersuccessful solo career of her own. (In fact, her album *Tapestry* still holds the record for the longest length of time a female artist has held the number-one spot on the Billboard charts.)[*]

Aside from King's influence on the Detroit sound, female songwriters were also dominating New York City's burgeoning folk scene in the swinging sixties. If Bob Dylan was the king of folk, then songstress **Joan Baez** was the queen. In addition to

---

[*] *Tapestry* remained at the top spot on the Billboard chart for fifteen consecutive weeks.

being Dylan's protest-song peer, she was his muse, having been said to inspire many of the romantic odes on albums like *Bringing It Back Home* and *Blonde on Blonde*. At the same time Baez was using music as a political podium, **Joni Mitchell** was honing her unique guitar style and inflective vocals in the West Village, proving that anyone with a message and a voice could join in the chorus of a new generation.

A couple years later, on the other side of Broadway, a different kind of musical revolution was starting to bubble to the surface—and these ladies were more concerned with being pissed off than politically correct. **Debbie Harry**, a onetime mild-mannered secretary, may've ditched her bouffant for a bottle of peroxide when she decided to front Blondie, but her mussed-up punk-rock prowess only made her more beautiful—and iconic. **Patti Smith**, on the other hand, was not one of the lookers, but her overwhelming stage presence and protagonist musical style made her utterly captivating, endlessly powerful, and oddly attractive. Then there's Sonic Youth frontwoman and bassist **Kim Gordon**, who added a much-needed dose of feminine mystique to NYC's No Wave scene in the early 1980s. In addition to being an early grunge goddess, Gordon can also add fashion designer, video director, and art curator to her lucrative résumé, making her one of the first music geek girl moguls.

Women weren't only affecting the shape of NYC punk to come; they were also changing the rock soundscape on the West Coast. In the late 1970s, heavily inspired by the glam-rock sound of David Bowie and the leather-clad look of **Suzi Quatro**, **Joan Jett** and the Runaways became the pouty-faced poster children for underage female rebellion. Whereas some girls were trolling the Sunset Strip looking for tail, Jett & Co. were looking to rock—and maybe bash a few skulls. It's also impossible to mention this era of the Los Angeles music scene without talking about X and frontwoman **Exene Cervenka**, who believed that

"The thing I found so fascinating about [Pretenders frontwoman] Chrissie Hynde when I was growing up was what I found attractive about men. She wasn't wearing pretty skirts and being a victim and talking about love. She was standing at the microphone with her legs spread, she was playing her guitar, and she was the coolest sight I'd ever seen in my life." —Shirley Manson, Garbage

punk was supposed to be both personal and provocative—a foreign concept for those who believed the genre was merely about Doc Martens, safety pins, and anarchy with a capital *A*.

Many of the aforementioned music geek godmothers acted as inspiration for the feminist musical revolution happening up in the Pacific Northwest during the early 1990s. Bands like Sleater-Kinney, Bikini Kill, and L7 ushered in a new era of independent-minded punk rock but also fostered an entire subculture of girls looking to expand their worldview musically, artistically, and politically. It wasn't enough to pick up a guitar and imitate your idols; girls were encouraged to speak out and challenge social norms. The riot grrrl ethos was similar to the lyrics of "Anything You Can Do, I Can Do Better" from the musical *Annie Get Your Gun*—although I don't think that was necessarily what **Kathleen Hanna** was thinking when she released her infamous manifesto in 1991.

"The future of rock belongs to women."
—Kurt Cobain, 1994

At the turn of the century, music mavens opted for a less in-your-face proclamation of their geekiness. With newsprint being replaced by TypePad and local meet-ups turning into on-line message boards, the revolution went digital. Finally, music geek girls all over the world could connect to, learn from, and interact with each other. If not for the blanket reach of the Internet, music-thirsty female fans might not have discovered **Feist**'s ambient indie rock or Metric's synth-tastic electro-pop. They might not have experienced the Distillers' polished punk anthems or whatever weird warbling **Joanna Newsom** passes off as folk music. In other words, for better or for Joanna Newsom, music geek girls have come a long way over the past fifty years. Luckily for us, they're just getting started.

COURTNEY LOVE: MUSIC GEEK OR MUSIC FREAK?

Much like olives, jeggings, and episodes of *Two and a Half Men*, when it comes to Courtney Love, you either love her or hate her; there's no in-between. However, regardless of how you feel about her freaky Renaissance Faire fashion sense, allegedly questionable parenting skills, and seemingly endless trips to court (and rehab), it's hard to deny that deep down, beneath the Stevie Nicks garage-sale garb and inevitable oxycodone residue, Courtney is a music geek icon.

Back in the late 1980s, it was her unhealthy obsession with post-punkers Echo & the Bunnymen that caused a then-sixteen-year-old Love to hop the Atlantic and invade Britain in search of New Wave grandeur. When that didn't pan out, she returned stateside, tried singing for Faith No More, got kicked out of Faith No More, hooked up with friend Kat Bjelland, joined riot grrrl upstarts

Babes in Toyland, got kicked out of Babes in Toyland, decided to start her own group called Hole, and eventually became the grunge goddess we know—and some of us love—today.

I don't care what anyone says: I think Courtney Love is one of music's most important geek girls. Sure, she's wacky and unbalanced, sometimes unintelligible and unnerving, but she always remains unwaveringly excited and passionate about one thing—the music.

MUSIC GEEK GODDESSES

If we don't take a look at where we came from, it's impossible to see where we're going. So let's give the preceding she-geeks a big round of applause, because if not for them breaking through the music scene's glass ceiling, we wouldn't be singing along with the following rock goddesses.

BJÖRK, Voted Avant-Garde and Ethereal Music Geek

I might not always understand what Björk's singing—or what she's wearing—but I've always appreciated her unwavering devotion to creating the most authentic and unusual music possible. An important aspect of being a music geek girl is trusting your gut. If your gut tells you to train with Inuit throat singers or wear a swan dress to the 2001 Academy Awards, well, you've gotta go with it. Part of what makes Björk a music goddess is her flair for dichotomy and her ability to remain mysterious yet misunderstood, powerful yet vulnerable. Björk is truly fearless, and that's an admirable quality in anyone—especially a music geek icon.

"For a person as obsessed with music as I am, I always hear a song in the back of my head, all the time, and that usually is my own tune. I've done that all my life." —Björk

CHAN MARSHALL, Voted Slightly Unbalanced and Overly Emotional Geek

Over the past fifteen years, Chan Marshall's musical minimalism has made her a touchstone for anyone who's ever been punched in the face by love. That sh*t hurts, right? But when the swelling goes down, the tears dry up, and it's time to properly nurse your wounds, there's no better soundtrack than the songs of Cat Power. If you've been lucky enough to see Cat Power perform—and I say this because Marshall has a tendency to cancel shows at the last minute or walk off mid-song without any explanation—then you know Marshall is one big ball of conflicting emotions that translates into some of the most heartfelt and heartbreaking indie rock ever recorded. Some people might call her psychotic; I like to think she's overly self-aware. Feel free to say the same if anyone tries to throw the crazy card your way, too!

JENNY LEWIS, Voted Nostalgic California Dreamin' Music Geek

Jenny Lewis has come a long way since singing about Wilderness Girls cookies in front of the Giorgio store in *Troop Beverly Hills*. (In case you don't remember, take a minute and search "Cookie Time" on YouTube. It's totally worth it.) In fact, in the early 1990s, it looked like this raven-haired powerhouse was destined for movie stardom, what with over twenty acting credits before the age of twenty. Thankfully, Jenny couldn't silence her love of music and left the bright lights of Hollywood for the urban sprawl of Omaha, where she joined the musical family at Saddle Creek Records and recorded with her band Rilo Kiley. Over the next fifteen years, Jenny rose up the indie-rock ranks to become one of the scene's most adored and admired frontwomen, thanks to her effortlessly cool style and relatability. Sure, she's still close with some of the A-listers from her past life,* but Jenny lives and breathes music, which you can tell by listening to any of her heartfelt and timeless recordings.

LIZ PHAIR, Voted Potty-Mouthed, Girl-Power Music Geek

When you're a teenager, hormones make everything more complicated. They make you fight with your parents, they encourage you to ditch third period, and they even spark an involuntary attraction to that weird guy in your English class who smells like scrambled eggs and wears long-sleeved button-downs underneath his T-shirts. Thankfully, when you're at your breaking point and don't understand why your insides are making a mess of the outside, there's Liz Phair and her arsenal of teen-angst an-

---

* For example, Lewis caused quite the stir when she accompanied pal Jake Gyllenhaal to the 2011 Golden Globe Awards. Don't worry, Taylor Swift. They went just as friends. Jenny is happily involved with fellow musician Jonathan Rice.

thems. Critics and music fans were blindsided when the Oberlin grad invaded the alt-rock scene in the early 1990s with revolutionary albums like *Exile in Guyville* and *Whip-Smart*. After all, it wasn't every day you heard a pretty and petite girl sing lyrics like "Every time I see your face, I think of things unpure unchaste, I want to f*** you like a dog, I'll take you home and make you like it."* Thanks to Phair's unapologetic sentiments about sexuality, crops of "flowers" have a better understanding of their bodies, themselves, and their healthy desire to be plucked.

> "A music geek girl is a girl who drowns her woes with Tegan & Sara instead of Ben & Jerry's."
>
> *Stephanie Ogozaly*
> *Carbondale, PA*

TEGAN & SARA, Voted Mind-Meld Twin-Sister Music Geeks

I don't know about you, but I feel like many of my music-geek kindred spirits are actually sisters from other misters. For that, I'm so jealous of Tegan and Sara Quin's musical—and biological—bond. (I wonder if they can read each other's mind like Mary-Kate and Ashley Olsen.†) Although the twins are barely thirty, they've been making beautiful music together since they were teens in their native Canada. Six critically acclaimed and fan-beloved albums later, Tegan & Sara have become pinups for indie-pop lovers everywhere—especially those who tend to read a lot of books, wear horizontal-striped T-shirts, watch reruns of *The L Word*, and cut their hair with a Flowbee.

> "This next song is about when you get your heart broken and you try your best to glue it back together and you wake up one morning and you're so happy because you realize, 'Oh my God, the tape's holding.'"
>
> —Sara Quin

HALL OF FAME: REGINA SPEKTOR

"Isn't it strange how we all feel a little bit weird?" Truer words were never spoken, Taylor Hanson, especially when it comes

---

* Taken from the song "Flower."
† This is sheer conjecture on my part, but if anyone's going to read minds, it's the Olsen twins.

to burgeoning baroque-pop* (and music geek girl) icon Regina Spektor. She started by playing dingy clubs in New York City's underground anti-folk scene but was drawn from obscurity when Julian Casablancas, the singer of the Strokes, handpicked her to open for them on tour. Although her music tends to take a more classical slant, her lyrical inspiration is all over the place (e.g., cancer, carbon monoxide, communism, and a bunch of other unexpected topics that don't start with the letter *c*.) Sure, she might've achieved mainstream success with songs like "Better" and "Us," which was the unofficial anthem of *(500) Days of Summer*, but Spektor will always be a musical misfit who isn't afraid to own her weirdness and wave her freak flag proudly, which only makes *us* worship her even more.

> "I'm the hero of the story, don't need to be saved."
> —Regina Spektor from "Hero of the Story"

MOVE ASIDE, WILLIAM MILLER

When people think about the field of rock journalism, most believe it's just like the movie *Almost Famous*—backstage parties, well-endowed groupies, and the overindulgent antics of spoiled musicians. Sure, all that stuff happens,† but it's secondary to the mission of any good rock writer, which is to go behind the music and bring the fan closer to his or her favorite band. Many have ventured down this boulevard of broken dreams, but few have made it out alive—or at least without a nasty addiction to cough syrup.‡ Let's take

---

* Sadly, I did not come up with the name for this genre. (It's pretty brilliant though, isn't it?) It's a term dating back to the 1960s and is used to describe the mash-up of chamber music with rock 'n' roll. Modern-day artists who would fall under this musical umbrella would include Tori Amos and Rufus Wainright.

† For example, during my rock writing days, I was on tour with a particular pop-punk band and rudely awoken on the bus at 3:00 a.m. to one of the band members screaming, "DUDE! Did you *really* just throw that used condom in my bunk? Where is it, dude? WHERE! IS! IT?!" Voms.

‡ R.I.P. Lester Bangs.

a minute to acknowledge the women who've managed to infiltrate this male-dominated profession and prove that if a girl's backstage at a concert, it's *not* because she's waiting to get her boobs signed.

Ann Powers, pop critic at the *Los Angeles Times*: Before taking her current post on the Left Coast, Powers spent time as the *New York Times* pop critic and a senior curator at the Experience Music Project. She's been hugely active in the female music writer community—as seen in the book she co-edited, *Rock She Wrote: Women Write About Rock, Pop, and Rap*—and can often be found appearing on panels or speaking at conferences on the topic of gender and music journalism.

Jancee Dunn, music writer and former host of MTV2's *120 Minutes*: Jancee spent most of the 1990s deep in the alt-rock trenches, first at *Rolling Stone* and then at MTV2. When I first read Jancee's pitch-perfect memoir, *But Enough About Me: How a Small-Town Girl Went from Shag Carpet to the Red Carpet*, I remember thinking, "Finally! Someone who gets my geeky musical neuroses!" Sure, I've never been French-kissed by Barry White or had the pleasure of sitting in Dolly Parton's kitchen, but Dunn's writing is so fantastic because she makes you *feel* like you're right there with her . . . and Barry . . . and Dolly. Talk about a threesome.

"I am thankful every day that I was born with the ability to hear, because I do not know how I would get through certain days without the music that I love."

Johanna Perez
Muncie, IN

Jane Scott, former rock critic at the Cleveland *Plain Dealer*: Not only is Scott widely considered the first female music writer, but many refer to her as the "World's Oldest Rock Critic." Her career took off after interviewing the Beatles back in 1964 and continued until 2002, when she retired from her post at the *Plain Dealer*. (She was in her early eighties!) Over her tenure at the paper, she attended nearly ten thousand concerts, interviewed countless rock stars, and was pretty influential in bringing the Rock and Roll Hall of Fame to Cleveland.

Sia Michel, pop music editor at the *New York Times*: After working at *Spin* for only five years, Sia took over as the magazine's editor in chief, making her the first woman to edit an American rock-music rag. While at the top of the masthead, she was intent on tearing the magazine away from the clutches of nu metal

"I don't know why anyone would do this if they didn't love it—the pay is generally horrific, but I still can't believe how lucky I am to have this job. There are times when I feel like a sh*thead for being honest about not liking something (I don't especially delight in writing negative reviews), and I wish that all writers were paid better (or at least on time). But I get to write about music all day long, which is the only thing I've ever wanted to do, and 99 percent of the time, that's just as fun as it sounds."
—rock critic Amanda Petrusich

and taking a chance on more up-and-coming bands.* The risk paid off and *Spin* was lauded as a music magazine with a soul. Unfortunately, Sia left the magazine in 2006 when *Spin* was bought out by new owners, but she didn't have to wait in the unemployment line too long. After a brief stint freelancing, she took over as pop music editor at the *New York Times*, a position she still holds today.

Vanessa Grigoriadis, contributing editor at *New York* magazine, *Rolling Stone,* and *Vanity Fair*: The only thing more impressive than Vanessa's résumé is the diversity of subjects she's been able to profile. From Olympic snowboarder Shaun White to genre bender Adam Lambert to sole man Christian Louboutin, Vanessa has spun tales about them all, which is what distinguishes her from all the other one-beat ponies out there—well, that and the enviable access she's granted. One read through her piece "Growing Up Gaga" for *New York* magazine, which outlined the rise of Lady Gaga from high school to arena stage, and you can practically hear the Monsters[†] squeal over the intimate confessions of the divine Miss G.

HONORABLE MENTIONS

Jenny Eliscu, contributing editor at *Rolling Stone*: While she may've penned *RS* cover stories on pop icons like Britney Spears and the Jonas Brothers, Jenny shows off her refined indie-rock tastes every day on SiriusXM satellite radio as a host on *The Spectrum* and *XMU.*

---

* Under Michel's editorship, *Spin* published the first U.S. covers of the White Stripes and the Strokes.
† The name of Lady Gaga fanatics.

Maura Johnston, music editor at the *Village Voice*: Always the unadulterated pop fanatic, Maura spent years establishing the voice and tone of Idolator.com before bringing her know-how to the classroom as an adjunct professor at NYU's Tisch School of the Arts. (She taught the course "Writing About Popular Music.") Currently, you can read Maura's musings in the pages of the *Village Voice*'s esteemed music section.

Amanda Petrusich, staff writer at Pitchfork.com: In addition to writing the pop music listings for the *New York Times*, Amanda also contributes to *Spin*, the *Village Voice*, and Pitchfork.com. No matter the outlet, she manages to balance objective critique with the personal aspects of songwriting and cites fellow critic Ann Powers as a major influence.

"OUCH, MY FOOT HURTS!"

Ah, Pitchfork. You hate because you love. Or, depending on the band, you really just hate because you hate.[*] The music scene would be so empty without your pithy reviews and somewhat elitist views on pop, rock, and everything in between. No band is safe on the site—unless you happen to be in a group like Do Make Say Think or Silver Jews. It seems like the more obtuse you are, the more stars you snag in the reviews section. But how many real people actually *listen* to those bands? How many music lovers think to themselves, "I absolutely *cannot* live another minute without listening to a Voxtrot record *right now!*"[†]

---

[*] Poor John Coltrane. If you were still alive to read the site's review of *Live at the Village Vanguard*, I'd feel really bad for you. Howev, you're dead, so I guess I feel bad for you either way.

[†] The correct answer is none, for those playing along at home.

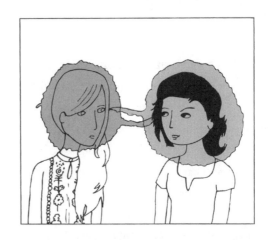

And so, should you ever come face-to-face with Ryan Schreiber—or someone equally as opinionated about all things music—you'll need to be prepared. Here's a crib sheet of the top ten bands you need to know in order to keep up with potential discourse, diatribes, and debates. Each is described in fifteen words or less because, let's face it, anything more and you might as well just read the long-winded review on Pitchfork.

Animal Collective: Pompous neo-psychedelia that sounds a lot better after you've eaten some pot brownies.

Deerhunter: Ambient indie pop with a twinge of Pavement-esque, jingle-jangle jams.

Dirty Projectors: Kooky indie rock where band members sound like they're playing different songs at the same time.

Grizzly Bear: Lo-fi psych folk that's equally inspired by Americana and Motown.

Isis: Yawn-inducing sludge rock that seems to excite both hipsters and metalheads.

Lightning Bolt: Noise disguised as music—or music disguised as noise? Your guess is as good as mine.

Mogwai: Atmospheric instrumental post-rock some people use instead of Ambien. In other words, sh*t's boring.

Mono: Shoe-gazey post-rock band from Japan who write ten-minute-long minimalist jams.

"I believe the biggest misconception about music geek girls is that we're all snobby hipsters. Yeah, I enjoy obscure music and like weird things, but that doesn't mean that I think I'm better than you for it."

Amanda Peschek
Milwaukee, WI

Slint: Math rock that adds up to sparse guitars and a singer who mumbles.

Tortoise: Post-punk meets cool jazz—aka the musical equivalent of watching paint dry.

FRENEMIES

Sure, our geeky musical muses have long since proved that girls belong *onstage* instead of just sidestage—or, as some sex-crazed guitar techs and misogynistic tour managers would prefer, "backstage,"* if you know what I mean—but music she-geeks still have a ways to go before they're treated like vinyl-collecting equals. In order to make true progress, our female audiophiles are going to have to tackle these frenemies first.

"They don't even know what it is to be a fan. Y'know? To truly love some silly little piece of music, or some band, so much that it hurts."
—Sapphire, in *Almost Famous*

✔ People who receive vote number reminders on their mobile phones.
✔ Bros who insist on invading college bars, requesting the O.A.R. song "That Was a Crazy Game of Poker," and proceed to loudly—and drunkenly—sing along while spilling their pitchers of Miller High Life all over anyone within a ten-foot radius.
✔ Claymates, Parrotheads, Phans,† or anyone who plans their vacation time around the Jam Cruise, the Rock Boat, or the Mayercraft.
✔ Anyone who listens to Insane Clown Posse and/or attends the annual Gathering of the Juggalos.

---

* Imagine me doing air quotes with my fingers while saying this to insinuate the whole point that lots of dumb dudes in the music scene still assume that all girls are groupies. These dudes are called "clueless asshats" or "sexist douche bags." (Again, more air quotes.)
† Nickname for Phish fans.

- ✔ Consumers who continue to buy those god-awful *Now That's What I Call Music* compilations.
- ✔ Miley Cyrus fans.
- ✔ Volvo-driving soccer moms who believe Green Day and Blink-182 invented punk rock.
- ✔ Any of the 658 people who bought Heidi Montag's debut album *Superficial* the first week it was released.

GEEK LOVE

When it comes to love, music geek girls want what everyone else does: a guy who worships their mind, body, soul, and CD collection—and then writes a tender love song about it.* Is that so much to ask? No. But is that pretty hard to find? You betcha. If our she-geeks have a refined taste when it comes to music, it's no wonder they're equally fussy when it comes to the perfect match.

A word of warning: there are lots of dudes to choose from in the music scene—whether it's the unassuming, long-haired clerk behind the counter at your local record store or the lanky, prepped-out frontman of your favorite indie-rock band—but male music geeks, in particular, are a tricky breed. They might look harmless, but they can be more heartbreaking than a Jeff Buckley song. It's for this reason that you must proceed with caution, because the only thing worse than falling for a guy who knows *nothing* about music is one who thinks he knows *everything*.

"What came first, the music or the misery? People worry about kids playing with guns, or watching violent videos, that some sort of culture of violence will take them over. Nobody worries about kids listening to thousands, literally thousands of songs about heartbreak, rejection, pain, misery, and loss. Did I listen to pop music because I was miserable? Or was I miserable because I listened to pop music?"
—Rob Gordon in *High Fidelity*

---

* One that I don't have to pretend to like because it's *actually good*.

"A couple of weeks ago, I overheard a guy in one of my classes mention that he thinks Jawbreaker is one of the best bands of all time. So, do I walk over and tell him that I love Jawbreaker, too? Impossible. Instead, last Monday I wore my Jawbreaker T-shirt. The aforementioned male flipped and hung back after class to talk to me and immediately friended me on Facebook. True love blossoming? Maybe. And all thanks to Hot Topic and the magic of the band tee."

Magdalena Burnham New York, NY

THE PERFECT MATCH FOR A MUSIC GEEK GIRL . . .

- ❏ Looks like Conor Oberst, Jack White, or Walter Schreifels.
- ❏ Doesn't mind going by the pet name "Plus One."
- ❏ Still makes CD mixtapes, complete with handmade cover art and thematic album titles.
- ❏ Remains optimistic that a reunion of Talking Heads or At the Drive-In *will happen* during his lifetime.
- ❏ Keeps past ticket stubs and music-related memorabilia in a photo album, shoebox, or drawer.
- ❏ Is always up for a show, even if he hasn't heard the band before.
- ❏ Collects vinyl but isn't an obsessive-compulsive wack-job who constantly brings up phrases like "test pressing," "picture disc," "vacuum tubes," or "shrink-wrap" in everyday conversation.
- ❏ Might be dissatisfied with the state of music today but vows to never become cynical about it.

REQUIRED READING

But Enough About Me: How a Small-Town Girl Went from Shag Carpet to the Red Carpet by Jancee Dunn

Girls to the Front: The True Story of the Riot Grrrl Revolution by Sara Marcus

Girls Rock: How to Get Your Group Together and Make Some Noise by Robyn Goodmark

Just Kids by Patti Smith

Let It Blurt: The Life and Times of Lester Bangs, America's Greatest Rock Critic by Jim Derogatis

Love Is a Mix Tape: Life and Loss, One Song at a Time by Rob Sheffield

Perfect from Now On: How Indie Rock Saved My Life by John Sellers

Please Kill Me: The Uncensored Oral History of Punk by Legs McNeil and Gillian McCain

Rock Bottom: Dark Moments in Music Babylon by Pamela Des Barres

This Band Could Be Your Life: Scenes from the American Indie Underground 1981–1991 by Michael Azzerad

Wish You Were Here: An Essential Guide to Your Favorite Music Scenes—From Punk to Indie and Everything in Between by Leslie Simon

"I would rather be at a concert on the barricade instead of out with a guy. Okay, it's not like I have guys beating down my door, but if a dude were to ask me out and I had New Found Glory tickets, you better believe I'd pick fist pumping over making out— unless he wanted to go to the show. Then, well, I don't know. I'll let you know when that happens."

Kendra Beltran
Cabazon, CA

WEB BOOKMARKS

AbsolutePunk.com

BrooklynVegan.com

Daytrotter.com

GorillavsBear.net

GrooveShark.com

Idolator.com

KCRW.com

NME.com

PasteMagazine.com

Pitchfork.com

PopDust.com

RCRDLBL.com

RollingStone.com

Spin.com

Spinner.com

Stereogum.com

MUST-SEE MOVIES

(500) Days of Summer

24 Hour Party People

Almost Famous

Control

High Fidelity

Joe Strummer: The Future Is Unwritten

Ladies and Gentlemen, The Fabulous Stains

Nick and Norah's Infinite Playlist

The Runaways

Sid & Nancy

Some Kind of Wonderful

This Is Spinal Tap

GEEK-APPROVED PLAYLIST

1. "Downtown Mayors Goodnight, Alley Kids Rule!" by You Say Party! We Say Die!
2. "The Police and The Private" by Metric
3. "Gold Lion" by Yeah Yeah Yeahs

4. "Sour Cherry" by the Kills
5. "Infinity Guitars" by Sleigh Bells
6. "Deceptacon" by Bikini Kill
7. "Volcano Girls" by Veruca Salt
8. "Connection" by Elastica
9. "Malibu" by Hole
10. "Silver Lining" by Rilo Kiley
11. "Back in Your Head" by Tegan & Sara
12. "Dance of the Seven Veils" by Liz Phair
13. "It's So Quiet" by Björk
14. "Not About Love" by Fiona Apple
15. "Here Is a Heart" by Jenny Owen Youngs
16. "One More Time with Feeling" by Regina Spektor
17. "The Greatest" by Cat Power

Five

FUNNY-GIRL GEEK

Not sure how you measure up in the witty world of funny-girl geeks? Ditch the canned laughter, grab a writing utensil, and test your she-geek skills!

1. Robin Williams is to Mrs. Doubtfire as Mindy Kaling is to:

    A. Ricky Gervais. Are you 'aving a laugh?!?

    B. Justin Bieber.

    C. Ben Affleck. Before this funny-girl geek made her big-screen debut in *The 40-Year-Old Virgin*, she portrayed Ben Affleck in the hilarious off-Broadway play *Matt & Ben*, a hysterical roast of Affleck and Damon's rise to fame after the script of *Good Will*

Hunting falls from the sky—or, more accurately, the ceiling of PS 122.

2. Which of the following is *not* an improvisational theater and training center for upcoming funny-girl geeks?

    A. Second City.

    B. The Groundlings.

    C. Franzia Station.

3. Actress Jane Krakowski plays self-centered fame-monger Jenna Maroney on NBC's *30 Rock*. However, in the series' original unaired pilot, a different comedienne played *The Girlie Show*'s female lead. Who was it?

    A. Amy Poehler.

    B. "Madonna. I remember seeing her *Girlie Show*
    world tour back in 1993 and she was fantastic. That
    was when vogueing was still an acceptable dance
    to do in public and before Madge morphed into the
    Crypt Keeper."

    C. Former *Saturday Night Live* mainstay Rachel
    Dratch originally played the part of Jenna DeCarlo.
    Executive producer Lorne Michaels later decided to
    replace Dratch with Krakowski and also change the
    last name of the character. Rachel still appears in
    the reshot version of the *30 Rock* pilot as a creepy,
    mannish cat wrangler.

4. Over the years, actress Elizabeth Banks has become quite the comedic muse for writer-director David Wain, who often casts the beauty alongside Wainy regulars like Paud Rudd, Ken Marino, and Joe Lo Truglio. Name the only Wain-directed film Banks has *not* starred in.

    A. "I don't know off the top of my head. Gimme a
    sec to pull up the IMDb app on my iPhone . . ."

B. "I didn't know that weird magician guy made movies, too. How does he have time to write screenplays when he's suspended in a Plexiglas box over the River Thames for forty-four days? Oh, wait . . . that's David *Blaine*? Then I have no idea who you're talking about."

C. *The Ten.*

5. You grab a soda from the fridge, but on your way to the living room, you accidentally drop it on the floor. *Whoopsie!* The can seems fine but when you place it on the coffee table and pull back the metal tab, the pop starts to fizz and erupt like a volcano. Do you think you can put your mouth on it to stop it from overflowing?

A. "Yes—at least until I run into the kitchen and grab a paper towel."

B. "No. I just applied my Lip Venom and I don't want to sabotage the gloss's plumping power."

C. "That's what she said."

ANSWER KEY

Mostly As: The effort is greatly appreciated but, unfortunately, you haven't read enough of *Dealbreaker: The Book for You, Man No Good* by Lesbian Yellow Sour Fruit to be considered a true funny-girl geek.

Mostly Bs: Sorry, Kim Kardashian, but we're going to have to bang the gong on this one. Plus, shouldn't you be busy eating Carl Jr.'s grilled chicken salad or having said salad extracted from your tush right now?

Mostly Cs: Congratulations, lady! You're a bona fide funny-girl geek!

CHARACTER SKETCH

Picture this: You're waiting in line at the movie theater and notice a tall, long-legged blonde positioned in front of you. She looks like a cross between Charlize Theron and Gisele Bundchen—but *prettier*[*]—and she's dressed like the cover model on this month's Anthropologie catalog. Impeccable from head to toe, she stands with her hand cocked on her hip, stares up at the marquee, and loudly sighs. "I just *can't* decide. . . . Should we see the new Reese Witherspoon romantic comedy or the action thriller with Vin Diesel?"

It's at this point you realize that she's not speaking to the room at large; instead, she's asking the mousy, nondescript girl that you hadn't noticed standing next to her until now. The mild-mannered wallflower shoves her hair behind her ears, adjusts the strap on her Flight of the Conchords "It's Business Time" tote bag, and, without skipping a beat, blurts out, "That's like asking me to choose between licking the bottom of a hobo's bare feet or washing my face with baby spit-up. It's a lose-lose situation."

Ladies and gentlemen, allow me to introduce you to the funny-girl geek. She might not stand out in a crowd, but look a little closer and you'll see someone who's the star of her own late-night comedy special.[†] This group of she-geeks doesn't have a general body type, but many tend to possess (or once in their youth possessed) some sort of physical challenge or genetic variation—for example, orthodontia, face freckles, or polydactylism[‡]—that influences and inspires their heightened sense of humor. Instead of wiping out on a wave of self-doubt and insecurity, the funny-girl geek would rather hang loose muttering

> "Mentally, she's sharper than a tack. Laughter is a funny-girl geek's drug of choice and let's hope for all of our sakes, she never encounters Dr. Drew—or rehab for it."
>
> *Estela Magana*
> *Orange, CA*

---

[*] As if that were humanly possible.

[†] Otherwise known as "her life."

[‡] A condition that leaves a human with extra fingers or toes—kind of like Count Tyrone Rugen in *The Princess Bride*.

one-liners under her breath. Anything for the sake of a good laugh—regardless of whether it's at her expense or the expense of others.

It is possible for funny-girl geeks to travel in like-minded packs; however, you have a better chance of finding them in one of two social scenarios: (1) as the token female in a group of funny-guy geeks, or (2) as the smart-ass sidekick to a girl who's a hot piece of eye candy (see above). Either way, the funny-girl geek is more than happy to play the part of the wisecracking accomplice and actually prefers to fly under the social radar. That way, she can carefully observe everyone's behavior, brainstorm astute non sequiturs, or simply mock incessantly.

It's for this reason that funny-girl geeks are often mistakenly branded as "mean," "spiteful," or "bitchy"—and that's not necessarily true. These sharp-tongued gals aren't out to intentionally hurt people, make them cry, or cause them to move to a different zip code because of the remote possibility of an impromptu run-in at the local Panera Bread. It's just that the tongue can be mightier than the sword, which is why so many funny-girl geeks are perceived as off-putting and intimidating. It's not personal, though. Trust me. Funny-girl geeks just love to laugh—whether it's *with* you or *at* you.

> "The biggest misconception about funny-girl geeks is that we never get hurt. Just because we make jokes about everything doesn't mean we're the emotional equivalent of Superman."
>
> *Alex Tracy*
> *Andover, CT*

" . . . BUT SHE HAS A GREAT PERSONALITY"

Herein lies one of life's most perplexing, existential, intricate, and intellectual arguments: can a girl be both pretty *and* funny? I'm not sure how the myth originated, but it seems like ever since the beginning of time, it was thought that a female could be witty or beautiful—never *both*.

Look at the legendary story about Helen of Troy, whose beauty incited actual battles and was later described as "the face that launched a thousand ships." She was kind of like the Megan Fox of her time—minus the painful acting skills and a gnarly case of clubbed thumbs. How much do you want to bet that the Trojan War would never have happened if Helen looked like, say, Janet Reno?

Sure, we've made a lot of progress since the Bronze Age, but we've got a long way to go. For example, why is it when a woman is described as having a "good personality," it somehow translates into her looking like Quasimodo's twin sister? Since when did having a "good sense of humor" become code for "Fire in the hole! We've got a grenade[*] here, people!"? Contrary to popular belief, beauty and brains are *not* contradictory traits. In fact, one cannot truly exist without the other. (Need proof? Two words: Carrie Prejean.) Thanks to fantastically sassy and unconventionally beautiful women like wisecracker Joan Rivers and social commentator Fran Lebowitz, here's to hoping that people won't be slinging around phrases like "sparkling personality" or "witty sense of humor" as insults for very much longer.

Hell, a girl can dream, can't she?

> "Good looks and a great personality are not mutually exclusive. One has no influence on the other, and to say that a pretty girl cannot be funny or that a funny girl cannot be pretty would be like comparing apples to basketballs. It just wouldn't make sense."
>
> *Amanda Lockhart*
> *Morehead, NC*

GEEK MYTHOLOGY

For many comedy purists, the she-geek lineage starts in the 1940s with comedienne-singer-actress **Fanny Brice**. Her depiction of Baby Snooks,[†] the mischievous make-believe toddler and title character of *The Baby Snooks Show* radio program, turned the vaudeville alumna into a star of stage and screen. In 1951,

---

[*] N. The ugliest girl in a gaggle of beauties. If a guy wishes to hook up with one of the group's prettier girls, then he has to make sure that the grenade gets action, too, so she doesn't feel left out and risk ruining his chances of going to Makeout City. Wingmen who take one for the team are often referred to as "grenade jumpers."

[†] Not to be confused with Nicole "Snooki" Polizzi from MTV's *Jersey Shore*.

Brice died from a cerebral hemorrhage at the age of fifty-nine, but her sass and charm continued to inspire future female farceurs, as exhibited in the 1968 musical *Funny Girl*, which was based on Brice's life and starred a then-unknown gawky, goofy ingénue named **Barbra Streisand.**

In the 1950s, Hollywood remained skeptical of a witty woman's ability to build a ticket-buying fan base because jaw-dropping starlets like Elizabeth Taylor and Grace Kelly were the belles du jour. However, offbeat charmers—like **Gracie Allen** (who played the comic yin to actor-husband George Burns's deadpan yang), **Joan Rivers** (who first introduced celebrity foi-

"Maybe pretty women weren't funny before because they had no reason to be funny. There was no point to it—people already liked you."
—humorist Patricia Marx

bles as a source of comedic fodder), **Phyllis Diller** (whom many critics consider the first female stand-up comedienne), and **Lucille Ball** (who mocked, and simultaneously celebrated, her own kookiness in TV's *I Love Lucy*)—refused to be chewed up and spit out by studio heads who were only interested in beauty, not brains.

Always one to inspire others and foster up-and-coming comediennes, it's no surprise that Ball went on to mentor one of the funniest females on television, **Carol Burnett**. Having spent much of her early career earning rave reviews on the Broadway and cabaret circuit, Burnett was offered her own sitcom by Ball's Desilu production company but turned the opportunity down in order to star in her own late-night variety show—something that was extremely risky at the time. But, in 1967, Burnett proved everyone wrong by launching *The Carol Burnett Show*, which ran for eleven years and garnered twenty-three Emmy Awards.

In the middle of *The Carol Burnett Show*'s run, a new type of sketch-comedy show debuted: one that was edgy, sharply satirical, and completely family-*un*friendly. That show was *Saturday Night Live*, and when it first aired in 1975, its original ensemble consisted of a motley crew of "not ready for prime-time players," including **Jane Curtin**, **Laraine Newman**, and **Gilda Radner**, who was actually the first person cast on the show. With her brilliant spoofs of Barbara Walters and Patti Smith, not to mention original characters like the brash big-banged Roseanne Roseannadanna and nasally congested über-nerd Lisa Loopner, Radner soon became the Queen of Parody—and without wearing a bra, no less (a fact that still stupefies me to this day). During the past thirty-five years, *SNL* has launched the sidesplitting careers of many funny-girl geeks—like **Julia Louis-Dreyfus** and **Joan Cusack**—even though some of them were only on the show for a short time, like **Janeane Garofalo**, **Laura Kightlinger**, and **Sarah Silverman**.

In recent years, the ever-evolving invention of the interweb and, more specifically, social-networking sites like Twitter, Facebook, and YouTube have made it easier than ever for funny-girl geeks to discover and connect with kindred spirits. First, there's **Aubrey Plaza**, who got her start as a deadpan drug-addled teen on the web series *The Jeannie Tate Show* and now appears as the deadpan intern April on NBC's *Parks and Recreation*. Then there's **Charlyne Yi**. Sure, Hollywood started to take the alt-comedy experimentalist seriously when she released her hybrid documentary *Paper Heart*, but most funny-girl fanatics celebrate and appreciate Yi for her awkward joke-telling tactics and interactive comedic approach. She's kind of like a modern-day Andy Kaufman—but female . . . and Asian. Finally, there's **Kate Micucci** and **Riki Lindhome**, who, separately, are both working actresses who have made small but memorable appearances on *Scrubs* and *Gilmore Girls*, respectively, but it's when the duo performs together as their guitar- and ukulele-wielding alter egos Garfunkel and Oates that they truly push comedic boundaries and manage to successfully infiltrate the male-dominated world of musical comedy. The two troubadours recently signed a development deal with HBO, and I hope they will be the first of many funny females who make their way from the stage to the screen.

> "I feel a little bit of pressure to be funny just because that's how people see me on TV. I definitely put that pressure on myself. I usually just come across weird and people think that's funny. I guess just being myself is funny, whether I mean for it to be or not." —actress-comedienne Aubrey Plaza

FUNNY-GIRL GEEK GODDESSES

Our geek godmothers worked tirelessly and endlessly to have their humor taken seriously by skeptics, and the benefits of their work have inarguably paid off for the newest crop of ridiculously witty women. Here are a few of my favorite modern-day funny-girl geek goddesses. Let the worshipping begin!

AMY POEHLER, Voted Funny-Girl Geek-Next-Door

Whenever I see Amy Poehler come onscreen, I laugh. With her seemingly wholesome and innocent demeanor, it doesn't matter what she's doing or what she's saying because I immediately start to giggle. The lady is *that* funny looking. Wait . . . that came out wrong. Amy Poehler started sharpening her funny bone while doing improv in college but really got serious about it when she moved to Chicago in the early 1990s and studied her craft at Second City and Improv Olympic. That's where she was introduced to Matt Besser, cofounder of the comedy troupe Upright Citizens Brigade, which she joined a couple years later. However, it was her time on *Saturday Night Live*—as co-anchor on "Weekend Update" and as characters like Netti Bo Dance from "Appalachian Emergency Room" and Amber, the one-legged girl with a 'tude (and gnarly sex addiction)—coupled with her recent work on *Parks and Recreation* that really solidify Amy's goddess status.

KRISTEN WIIG, Voted Careful or Your Face Will Freeze Like That Funny-Girl Geek

Aside from the occasional *SNL Digital Short* ("Man, this ain't my dad! This is a cell phone!"), Kristen Wiig just might be the best thing about *Saturday Night Live* these days. When it comes to her sense of humor, quiet is the new loud and Kristen is the queen of nonverbal communication. You don't have to do fart-wheels while singing "I'm Too Sexy" at the top of your lungs in order to make an impression; hilarity is just as powerful when it's delivered through a subtle gesture or an intense stare. Though it's not her *best* look (mostly because her chin instantly disappears and she morphs into the female version of Beeker from the Muppets), Kristen's trademark frown-pout is an automatic giggle-inducer, not to mention scene-stealer.

AMY SEDARIS, Voted Don't Know Whether to Laugh or Vomit Funny-Girl Geek

Much like Vegemite and Josh Groban, Amy Sedaris is an acquired taste. It's not because she's unfunny—it's almost that she's *too* funny. Her comedy knows no limits or boundaries, which can at times annoy people to no end or make them violently uncomfortable and even ill. Whether she's playing yellow-toothed former junkie and teenage runaway Jerri Blank on *Strangers with Candy* or ill-advising nesting fools on how to remove urine stains from the sheets in her book *I Like You: Hospitality Under the Influence*, Sedaris refuses to break character. *Ever.* Sometimes, I wonder if her entire life is one big inside joke between her and her rabbit Dusty. Let's just say nothing Amy does would surprise me—and that's why she's such a funny-girl geek goddess. Well, that and her mind-blowing relationship with Glen Fandango, her imaginary husband.

"Crafting, or 'making things,' has always been a delightful pastime of mine because it requires putting common elements together in order to achieve a lovely something that nobody needs."
—actress-comedienne-author Amy Sedaris

MINDY KALING, Voted Highbrow Funny-Girl Geek

Dartmouth grad Mindy Kaling first earned comedy raves for cowriting the critically acclaimed off-Broadway play *Matt & Ben*, which parodied the outlandish trajectory of acting buddies Matt Damon and Ben Affleck after the script for *Good Will Hunting* literally falls from the sky and hits them in the head. From there, Mindy started writing for *The Office* and was later added to the Dunder Mifflin Paper Company team as Kelly Kapoor, the obnoxiously loquacious customer service rep. Her undeniable intellect and high-brow wisecracks, though somewhat shaded by her Valley Girl delivery, helped define the show's signature voice and resulted in fan-favorite episodes like "Ben Franklin" and "Niagara." It's no wonder *Entertainment Weekly* voted her as one of 2009's "10 Funniest Actresses in Hollywood," and NBC recently signed her to a seven-figure contract to develop a new comedy for the channel. That sh*t is bananas! B-a-n-a-n-a-s!

CHELSEA HANDLER, Voted Oftentimes Drunk and/or Horizontal Funny-Girl Geek

You may've missed Chelsea when she appeared in the Oxygen Network female-centric prank show *Girls Behaving Badly*,* but it's this comedienne's legendary bad behavior that's taken her out of open-mic-night hell and into late-night talk show heaven. A lot of naysayers scoffed at the thought of a funny female (wo)manning her own show, but loudmouthed Handler got the last laugh when *Chelsea Lately* became one of E! Network's highest-rated shows. (Suck on that, Seacrest.) Oh, it also doesn't hurt that Chelsea's a *New York Times* best-selling author of such hilarious tomes as *My Horizontal*

---

* Don't worry. You're not the only one.

Life and *Lies Chelsea Told Me*, which is the first title from Handler's own imprint with Hachette Book Group. Don't worry. She's using her power as an opportunity to highlight up-and-coming comedic talent: the second release is a book by her dog, Chunk.

HALL OF FAME: TINA FEY

So who's the funniest she-geek of all? I'll have to go with Tina Fey for $1,000, Alex. *Duh.* Her unique humor has earned her endless awards and accolades—though she has yet to achieve EGOT status[*]—but I think her most impressive title is Fabulously Brilliant Unofficial Funny-Girl Geek Spokeswoman of the World and Beyond. No, I did not make that up.[†]

That's not to say the road from Second City improv diva to *Saturday Night Live* head writer to beloved creator and star of *30 Rock* has been paved with Cheesy Blasters. Like most funny-girl geeks, Fey grew up somewhat of a social misfit who spent much of her free time mocking and ridiculing those high school dummies who reveled in prom and parties instead of, say, actually getting invited to one or the other. Fey has also said that she was never a big hit with the guys, admitting on *The Late Show with David Letterman* that she "couldn't *give* it away back in the day," which is why she remained a virgin until the ripe old age of twenty-four.[‡] However, what she lacked in popularity and sexual prowess, she made up for in drive, motivation, and whip-smart humor.

If I may be so bold, I think that Tina Fey has successfully won the heart of the nation and redefined what it means to be

---

[*] A rare distinction for the small group of entertainers who have won an Emmy, Grammy, Oscar, *and* Tony Award. Previous EGOT winners include Audrey Hepburn, Mel Brooks, and Whoopi Goldberg, if you can believe it.

[†] Yes, I did.

[‡] Although not voluntarily.

"I want to thank my parents for somehow raising me to have confidence that is disproportionate with my looks and abilities. Well done. That's what all parents should do!"
—Tina Fey, in her acceptance speech for Outstanding Lead Actress in a Comedy Series for *30 Rock* at the 2008 Primetime Emmy Awards

America's Sweetheart—and all without ever having to star in a romantic comedy with Tom Hanks *or* Billy Crystal.

Sorry, Meg Ryan, but it looks like there's a new sheriff in town.

TINA FEY PAS

It's scientifically proven that there's no girl funnier than Tina "I Want to Go to There" Fey, especially when she's playing lovable loser Liz Lemon on *30 Rock*. Whether she's dishing out deal-breakers ("Long distance is the wrong distance!") or waxing poetic on organized religion ("I pretty much do whatever Oprah tells me to"), whenever she opens her mouth, catchphrases fall out. Want to know how you can coin your own snowclone?* It's actually easier than you think. Just follow a few easy steps and *voilà!* Instant idioms!

1. Take an old catchphrase and replace one of the words with a random item often found at a garage sale.
   "Where's the <u>beef</u>?" becomes "Where's the <u>hairnet</u>?"
   "You deserve a <u>break</u> today" becomes "You deserve a <u>Barbie sports car</u> today."
   "Yo quiero <u>Taco Bell</u>" becomes "Yo quiero <u>London Fog chaqueta</u>."

2. When you're tempted to use an expletive to express pain, surprise, or uncontainable joy, substitute the non-sensical name of one of Ikea's furniture series.
   "Son of a <u>bitch</u>!" becomes "Son of a <u>bestå</u>!"
   "You just hit my car, motherf<u>***er</u>!" becomes "You just hit my car, mother<u>flört</u>!"
   "*[Stubs toe.]* Holy <u>ass</u>!" becomes "*[Stubs toe.]* Holy <u>årstid</u>!"

3. If you're ever stuck in an uncomfortable conversation or social setting, start talking about your favorite appetizer, sandwich, cupcake, or combination of all three.
   Person: "Have you ever been to Italy?"
   You: "No, but if the Tuscan spinach dip at T.G.I. Friday's is at all representative of the country, I think I'd love it there."

---

* (This is a real word. Seriously.) N. Type of cliché or phrasal template. For example, "Snowclones are the new catchphrases."

Person: "I just read this really interesting article in *Mother Jones* about hyperconsuming. Did you happen to see it?"

You: "I didn't, but you know what I always say: Waste not, *won*ton tacos."

Jon Hamm: "Hi, my name is Jon Hamm."

You: "Nice to meet you! *[Pauses and steps back.]* Wait . . . didn't one of your relatives murder Mama Cass?"

4. When a guy hits on you, awkwardly twirl your hair around your finger, giggle nervously, and respond with the biggest overshare you can think of.

Guy: "Was that an earthquake or did you just rock my world?"

You: "Actually, I think I just let out a bottom burp. Sorry 'bout that."

Guy: "Let's make like a fabric softener and snuggle."

You: "Sounds great, but I should probably warn you that the eczema is flaring up on my back, so try not to nestle too close to any area that appears to be scaly, crusty, or oozing. Your place or mine?"

Guy: "I'm like chocolate pudding, I look like crap but I'm as sweet as can be."

You: "Really? Because I did the Master Cleanse once, and my crap looked more like rope than chocolate pudding. *[Rustles around in purse.]* I think I have some pictures saved on my phone. Wanna see?"

"I got a lovely compliment once. I had said something delightfully sarcastic and a girl that I barely knew cocked her head and said to me, 'You're so . . . interesting. I never know if you're kidding or not. You've got a unique personality.' I chose to accept it as a compliment and not dig deeper into its true meaning, which I assume was, 'You are weird and that's why we don't invite you to drink beer with us in the woods.'"

Stephanie Ogozaly
Carbondale, PA

5. If you ever get bored with a conversation, take whatever the other person last said and insert it after, "Your face . . ."

    Person: "Wow, those pot stickers look delicious!"
    You: "Your face is a pot sticker."

    Person: "I'm not sure what movie I wanna watch on Netflix. It's between *The Virgin Suicides* and *Lost in Translation*."
    You: "Your face is *Lost in Translation*."

    Person: "What do you think about the U.S. House of Representatives vote to bar Planned Parenthood from all federal funding? If the budget cut goes through, it could hugely affect women's access to HIV tests, cancer screenings, and contraceptives."
    You: "Your face is a contraceptive."

FRENEMIES

Thanks to the outstanding (and hilarious) efforts of our funny-girl goddesses, being described as "having a good sense of humor" doesn't have to be viewed as an insult. However, there are still those who don't understand what makes a funny-girl geek so . . . well . . . *funny*. In other words, if you're a disciple of any of the lovely ladies listed in this chapter, prepare to face some possible opposition—and blank stares—from the following group of frenemies.

- ✔ Investment bankers, stock brokers, and various other Wall Street douche bags.
- ✔ Girls who wear leggings instead of pants.
- ✔ Fans of Ed Hardy, Smet, or Affliction clothing.
- ✔ Spoilsports, wet blankets, party poopers, and general Debbie Downer types.
- ✔ People who think Dane Cook is funny.
- ✔ Women who wear fragrances by celebrities like Paris Hilton, Britney Spears, Jennifer Lopez, or anyone else who either (1) appears on the front page of TMZ.com on a daily basis or (2) has a reality show on E! Entertainment.* (Sorry, Kimora Lee Simmons, but that means you, too.)
- ✔ Sufferers of aphonogelia.†
- ✔ Dane Cook.

GEEK LOVE

Love is no laughing matter for the funny-girl geek. In fact, it takes a very special type of guy to appreciate—and understand—her quirk, quips, and odd physical attraction to Zach Galifianakis. It's no surprise, then, that most funny-girl geeks find romantic solace with similar comedic counterparts. Whether these guys can't go a day without quoting Will Ferrell—even lines from his crappier movies‡—or constantly debate which is better, the

---

* There are two, and only two, exceptions to this rule: Gwen Stefani's L.A.M.B. and Debbie Gibson's Electric Youth. Why? Because I said so. Natch!

† N. The neurological inability to laugh out loud. Kind of gives new meaning to the phrase, "So funny I forgot to laugh."

‡ Dear Will, I would really appreciate it if you could reimburse me for *Blades of Glory, Land of the Lost,* and *Bewitched*. Those movies stank more than Khloe Kardashian and Lamar Odom's unisex perfume. Please send a check or money order c/o my publisher. Thanks in advance. P.S.: Can you start making funny movies again? That'd be awesome. Best, Leslie.

British or the American version of *The Office*, they are the Will Arnett to our Amy Poehler.

Still not sure who's the right guy to fill up the funny-girl geek's love tank? The following cheat sheet should help narrow down the search.

THE PERFECT MATCH FOR A FUNNY-GIRL GEEK . . .

- ❏ Looks like Jimmy Fallon, Andy Samberg, or Michael Cera.
- ❏ Gets most of his news and political attitude from watching *The Daily Show* and/or *The Colbert Report*.
- ❏ Thinks Raaaaaaaandy, Aziz Ansari's comedic alter ego, is funnier than Aziz himself.
- ❏ Vows to never take a date for a bucket o' grub at Dick's Last Resort.
- ❏ Understands that being the life of the party doesn't always mean being the center of attention.
- ❏ Still mourns the loss of *Arrested Development*.

"At some point during almost every romantic comedy, the female lead suddenly trips and falls, stumbling helplessly over something ridiculous like a leaf, and then some Matthew McConaughey type either whips around the corner just in the nick of time to save her or is clumsily pulled down along with her. That event predictably leads to the magical moment of their first kiss. Please. I fall ALL the time. You know who comes and gets me? The bouncer."
—comedienne Chelsea Handler

- ❑ Has no idea who Robert Pattinson is.
- ❑ Doesn't find Scarlett Johansson, Jessica Alba, or Keira Knightley at *all* attractive—or at least never admits to it.

JUDD APATOW IS A PIMP

Although he might not be as dreamy as, say, Neil Hamburger, writer-director-producer Judd Apatow is quite the catch for any funny-girl geek. (Too bad he's already taken by actress Leslie Mann, a great example of the coexistence of brains and beauty.) In addition to creating cult-classic, geek-approved TV series like *Undeclared* and, of course, *Freaks and Geeks*, Apatow has also managed to redefine the modern-day romantic comedy with films like *The 40-Year-Old Virgin*, *Knocked Up,* and *Funny People.*

All of Apatow's films seem to cast paunchy (Seth Rogen pre–*The Green Hornet* weight loss), balding (sorry, Steve Carell, but it's true), and objectively unattractive (cover your eyes—Adam Sandler's coming!) men as the romantic lead, which is revolutionary when you stop to think about it. Take, for example, someone like Jonah Hill, who was recently cast to play opposite the succulent piece of eye candy known as Brad Pitt in the movie *Moneyball.* Doesn't that blow your mind a little? It certainly blows mine.* Just like Tina Fey is helping to take the title of America's Sweetheart into the twenty-first century, Apatow is single-handedly creating a new brand of male sex symbol—one who's slightly overweight, tends to sweat a lot, has hair kinkier than a bag of Arby's curly fries, and worships Steve Martin. In other words, the perfect potential husband for any funny-girl geek.

---

* Watch your step. I'd hate for you to slip and fall on my brains.

REQUIRED READING

Are You There, Vodka? It's Me, Chelsea by Chelsea Handler

The Bedwetter: Stories of Courage, Redemption, and Pee by
  Sarah Silverman

Bossypants by Tina Fey

*Feel This Book: An Essential Guide to Self-Empowerment,
  Spiritual Supremacy, and Sexual Satisfaction* by Ben Stiller
  and Janeane Garafalo

Hospitality Under the Influence by Amy Sedaris

How to Talk Dirty and Influence People by Lenny Bruce

Is Everyone Hanging Out Without Me? (and Other Concerns)
  by Mindy Kaling

Live from New York: An Uncensored History of Saturday
  Night Live, *As Told by Its Stars, Writers and Guests* by
  James A. Miller and Tom Shales

My Booky Wook: A Memoir of Sex, Drugs and Stand-Up by
  Russell Brand

The Will to Whatevs by Eugene Mirman

You're a Horrible Person, But I Like You: The Believer *Book
  of Advice* by Sarah Silverman, Zach Galifianakis, Fred
  Armisen, Judd Apatow, and others

WEB BOOKMARKS

BuzzFeed.com

CollegeHumor.com

Cracked.com

ComedyCentral.com

FunnyOrDie.com

Groundlings.com

ImprovEverywhere.com

Jezebel.com

MentalFloss.com
SecondCity.com
TheOnion.com
UCBTheatre.com

MUST-SEE TV

30 Rock
Absolutely Fabulous
Arrested Development
Community
Extras
Flight of the Conchords
It's Always Sunny in Philadelphia
The League
Little Britain
Michael & Michael Have Issues
The Mighty Boosh
The Office*
Parks and Recreation
Party Down
Reno 911!
Scrubs
The State
Tim and Eric Awesome Show, Great Job!

GEEK-APPROVED ALBUMS

Flight of the Conchords by Flight of the Conchords
Incredibad by the Lonely Island

---

* Both the American and the British versions.

Intimate Moments for a Sensual Evening by Aziz Ansari

Jesus Is Magic by Sarah Silverman

Laugh Out Lord by Neil Hamburger

Mitch All Together by Mitch Hedberg

Music Songs by Garfunkel and Oates

Party by Nick Swardson

Revolution by Margaret Cho

*Shut Up, You F***ing Baby!* by David Cross

Words, Words, Words by Bo Burnham

You Broke My Heart in 17 Places by Tracey Ullman

Six

DOMESTIC
GODDESS GEEK

Are you ever boggled by the difference between knitting and crocheting?* Release your grip on your Speed Stix needles, grab a Sharpie, and test your domestic goddess geek girl skills!

1. What is the definition of "feng shui"?

---

* "Knitting" is creating a piece of needlework by looping one continuous yarn, whereas "crocheting" is creating a piece of needlework by looping thread—one loop at a time—with one hooked needle instead of two long-shafted knitting needles. If you still don't understand, go ask your grandmother . . . or Julia Roberts. That lady is *hooked* on needlework. (*Get it?* Sometimes I leave myself in *stitches*. There I go again! Maybe it's time for me to take this routine on the road. Someone call my agent!)

A. I think I just saw something about that on *CBS: The Other America*. Does it have something to do with the Financial Executive Networking Group—or FENG, as they're often called?

B. It means "to pop a stiffy" in Chinese. 党,韦恩!*

C. It's an ancient Chinese discipline using the laws of Heaven and Earth to improve one's life and increase positivity by reorganizing one's surrounding aesthetics. In English, the phrase translates into "wind-water"; today, many call it the "art of flow." Feng shui is most often associated with interior design; in lay terms, it consists of cleaning up, arranging furniture, and incorporating your birthday symbol.

2. DMC embroidery floss is to macramé as Modge Podge is to:

A. Bone carving.

B. Gaining weight. Lay off the KFC Double Downs, tubby. *Geez.*

C. Decoupage.

3. Which crafts-gone-wrong website boasts the tagline "Where DIY meets WTF"?

A. DontYouLoveMyCrochetChessSetsy.com.

B. YourMacaroniNecklaceMakesMyStomach Upsetsy.com.

C. Regretsy.com.

4. Ever since chef Rachael Ray started her *30-Minute Meals* show on the Food Network in 2001, she's been slowly cultivating her own vernacular. "Sammie" is short for "sandwich," "stoup" is the combination of "stew" and "soup," etc. How-

---

* That means "Party on, Wayne" in Chinese. Yeah, I'm multilingual. No biggie.

ever, the rest of the world seemed to catch on in 2006 when the *Oxford American College Dictionary* decided to induct one of her favorite sayings into their next edition. What was the abbreviation?

> A. WWOD: what would Oprah do?
>
> B. TATO: there's always takeout.
>
> C. EVOO: extra-virgin olive oil.

5. At what temperature is a glue gun ready to bond materials together?

> A. 98.6 degrees.
>
> B. It depends on the temperature of the materials involved.
>
> C. 380 degrees.

ANSWER KEY

Mostly As: When it comes to enthusiasm, I give you two green thumbs up. However, you still have some work to do before I'll let you knit me a sweater or rearrange my furniture.

Mostly Bs: Sorry, Paris Hilton, but a dream catcher is a hand-decorated willow hoop intended to snare dreams, not a brand of synthetic hair extensions.* Instead of inevitably burning yourself with a glue gun, I suggest you stick to what you do best. *Um* . . . remind me—and the rest of the world—what that is again?

Mostly Cs: Standing ovation, design diva! You're a bona fide domestic goddess geek!

---

* You guffaw, but Paris Hilton actually has a signature line of hair extensions called DreamCatchers. I can only *dream* of the things that get caught in that web of fake hair.

CHARACTER SKETCH

Our imaginary camera zooms in on a girl in a clothing boutique. She's on the hunt for a dress to wear to a friend's upcoming engagement party. It's summertime, so she wants something easy, breezy, and relatively inexpensive. After fondling almost every garment in the store, she happens upon the most adorable retro-looking poplin halter dress, complete with a full skirt, nipped waist, and sweetheart neckline. Her heart starts to race as she turns over the tag to reveal the price, which is a heartbreaking $350. Her stomach drops, her shoulders slump, and, after realizing that there's no way she can afford the dream dress on her limited budget, she exits the store empty-handed and utterly defeated.

"Crafters don't grow old; they just get more bazaar."
—Unknown

Blinded by images of Joan Holloway, our girl failed to sight another female customer drawn to the same dress, rubbing the fabric between her fingers, taking measurements, and sketching a rough pattern. Girl No. 2 was *also* shopping on a meager budget for formalwear to don for a friend's engagement party.* However, instead of abandoning the perfect dress simply because of pesky financial constraints, she has decided to take matters into her own hands and make a one-of-a-kind version of the frock. This sew resourceful diva is, of course, our domestic goddess geek!

Much like the other sects of she-geeks we've met, this particular group thrives on being original and unique. But what sets a decor-inclined sister apart from the pack is her desire to express her distinctiveness any chance she gets. Domestic goddess geeks are often easy to spot because everything about them—their wardrobe, their eating habits, their obsession with Method cleaning products—declares, "I am the queen of my style! Feel free to curtsy in my presence."

---

* Why does it feel like everyone and their mother are getting engaged? Seriously. If I go to *one more* bridal shower, I'm going to make a toilet paper noose to hang myself with. How's that for a party game?

Many outsiders mistakenly conflate being a domestic goddess with being a housewife. The two are neither synonymous nor mutually exclusive. A housewife *can* be a domestic goddess,* but a domestic goddess doesn't necessarily *have* to be a housewife.

Whether or not you've got a spouse and a bunch of rug rats running around, getting in touch with your domestic self can be a somewhat solitary endeavor. If you're on the shy side, you might be happy as a clam to sit endlessly at the Singer sewing machine in your bedroom and stitch together a knee-length jersey skirt, not realizing that you haven't spoken to another human being in over two days. However, for some of our more loquacious sisters, things like stitching circles and craft groups† can provide a much-needed social connection to the like-minded outside world. Nothing sparks a true friendship better than bonding over embroidery thread and stitching together bracelets to celebrate newfound bestiehood!

GEEK MYTHOLOGY

In one form or another, domestic goddess geeks have existed since the beginning of time. Epigraphists might disagree, but I'm pretty sure hieroglyphics were one of the earliest forms of wallpaper. Just sayin'. Fast-forward a few millennia to the early 1900s and you'll meet **Elsie de Wolfe**, a noted London and New York City socialite who penned the influential book *The House in Good Taste* and is prominently known as the First Lady of

"I'm definitely not the type of girl to marry Joe Schmoe and be barefoot and pregnant, hustling around a kitchen for the rest of my life. That is the *complete* opposite of me! However, I do think it's important for a chick to be able to feed, fend, and fight for herself out in this crazy, big world. Home is a huge place for me. I love to have a comfort zone and I can't imagine *not* having a chill place to rest my head, watch *True Blood* reruns, and eat more than the occasional tub of Ben & Jerry's."

Cara Ann Rob
Tulsa, OK

---

* I'm looking at you, Donna Reed. Or, for those of you who need a real-life example, I'd probably cite someone like Kelly Ripa on the basis of her domestically intoxicating Electrolux appliance commercials alone. Plus, who wouldn't want to make a house pretty for a delicious dish like Mark Consuelos? Yum-my!
† For example, the Church of Craft is a great organization that unites budding seamstresses all over the country, and the Rock Paper Scissors Collective in Oakland, California, offers classes on everything from silk-screening to zine-making, all of which are run by local volunteers.

Interior Design. Her refreshing Victorian makeovers of Upper East Side abodes were more influential than the fashions on an episode of *Gossip Girl*. **Dorothy Draper** is another early design inspiration. Though she initially got her start in the male-dominated field of construction, Draper was also responsible for setting the decorating tone of Manhattan's high society. She, too, wrote a home-improvement handbook entitled *Decorating Is Fun! How to Be Your Own Decorator*, and was known for her bold approach and grandiose aesthetic.

New York City was not only the epicenter of female-driven interior design but also the backdrop for etiquette expert **Emily Post** and her unprecedented approach to propriety. Her 1922 manners-driven masterpiece, *Etiquette in Society, in Business, in Politics and at Home*, has been reprinted beaucoup times and remains a classic to this day—even though it sometimes seems like good manners have gone the way of the dinosaur. In fact, I could rattle off the names of about a dozen people I'd like to sentence to ninety days at the Emily Post Institute in Burlington, Vermont—and that's just off the top of my head . . .

While Emily Post was schooling people on the art of table manners, the incomparable **Julia Child** was setting said table with the most savory meals this side of the Atlantic. Most food-obsessed goddesses consider Ms. Child the gold standard of cooking—whether they became smitten with her book *The Art of French Cooking* or the 2009 chick-flick *Julie and Julia*, which is based on her life in France during the time she wrote her groundbreaking cookbook. Either way, her influence on the modern female food movement is invaluable. Without Julia Child, we wouldn't have present-day kitchen queens like **Ina Garten, Paula Deen**, or **Rachael Ray**. And then there is Child's greatest protégé—Martha.

Martha Stewart might be the ultimate domestic diva, but she better look out for her throne. There are a bunch of burgeoning goddesses who are making a name for themselves and inspiring

> "There is no reason why you should be bored when you can be otherwise. But if you find yourself sitting in the hedgerow with nothing but weeds, there is no reason for shutting your eyes and seeing nothing, instead of finding what beauty you may in the weeds. To put it cynically, life is too short to waste it in drawing blanks. Therefore, it is up to you to find as many pictures to put on your blank pages as possible."
> —manners guru Emily Post

a new generation of craft-loving design geekettes. For example, there's **Wendy Mullin,** founder of the Americana clothing line Built by Wendy and author of the book *Sew U*, which offers step-by-step instructions on sewing, patternmaking, and clothing construction, but with a hipster twist. Then there's craftivist **Faythe Levine,** who captured the alt-craft scene in her documentary *Handmade Nation: The Rise of DIY Art, Craft, and Design*. Plus, let's not forget about star knitter **Vickie Howell,**

"Cooking is like love; it should be entered into with abandon or not at all." —chef Julia Child

who has taken needle arts to the next level by hosting HGTV's *Knitty Gritty* for eight seasons and writing a celebrity column for *Vogue*'s knitting magazine *Knit.1*, in addition to forming Craft Corps, an inspiring movement that celebrates the craft community. In fact, thanks to the Internet and the new ease of finding like-minded mavens, there's never been a better time to start getting crafty.

DOMESTIC GODDESS GEEKS

The domestic goddess geeks of yore have paved the way for the next generation of alt-craft mavens, food-loving females, and home-decor divas. In fact, if you want to get technical, they also put down the gravel, spread the asphalt, and drove the drum roller to seal in the pavement. Yes, ma'am, those old broads could do things with a bottle cap, a chocolate bar, and a piece of duct tape that would make even MacGyver blush. Inspired by hard work, a lot of elbow grease, and an unwavering passion for DIY design, the latest crop of goddesses continue to challenge conventional aesthetics, buck tradition, and create beautiful— and delicious—things. More important, they've helped turn domesticity from a household chore into a true art form. Brava!

CHRISTINA BATCH-LEE, Voted DIY or Die Domestic Goddess Geek

Christina Batch-Lee might not be a household name, but thanks to her unconventional marketing skills and passion for painting, she's helped turn parent company Etsy.com into one. Before joining the Etsy team, Batch-Lee started her own alt-craft magazine called *Adorn*, a periodical dedicated to all the unique fashion, travel, and interior design needs of the crafty girl-on-the-go. Christina is a walking example of Etsy's mantra "Buy,

"[The domestic goddess geek] is unashamed of how mom-ish her hobbies are, and she and her like-minded friends probably have a sewing or knitting circle every Thursday night. When you ask her where she got that awesome dress she's wearing, she tells you she made it out of old socks and quickly describes how, causing you to go home, try to imitate her handiwork, and end up crying over four pairs of chopped-up socks and cursing yourself for not having her magical abilities."

Katie Wright
New York, NY

Sell, Live Handmade," and her enthusiasm for alt crafts is both infectious and inspiring.

GENEVIEVE GORDER, Voted Design Problem-Solver Domestic Goddess Geek

We first became smitten with Genevieve Gorder when she appeared on everyone's favorite flip-flop decorating show *Trading Spaces*; however, we fell into flawlessly designed love when she launched her own show on HGTV, *Dear Genevieve*, where she tackles all sorts of real-life design challenges. She's kind of like the Ann Landers of home decor. Unlike some decorators who come up with grandiose—read: très expensive—ideas to give your space a facelift, Genevieve's advice is realistic and oftentimes inexpensive. She proves that no problem is too big or too small and you don't need a nod from the National Council for Interior Design Qualification in order to solve a design dilemma.

JENNY HART, Voted Oh Sew Talented Domestic Goddess Geek

As part of the infamous Austin Craft Mafia, Jenny Hart and her company Sublime Stitching have brought embroidery out of the pages of the L.L. Bean catalog and into the hands of crafty divas everywhere. Her revolutionary approach to the handicraft pairs traditional needlework with alternative designs and edgy patterns. (For example, transfer bundles include themes like "Gnomes and Fairies," "Viva Las Vegas," and "Om Sweet Om.") Jenny even packages together a Stitch-It Kit that includes all the materials needed to take a crack at cross-stitching plus an easy-to-follow illustrated instruction booklet that breaks down each stitch. Cheers to you, Jenny Hart, for giving my fellow alt-craft dames and me the itch to stitch!

"[Domestic goddess geek girls] can't NOT be working on something. Whether it's crocheting an adorable hat, knitting a warm and cozy scarf, or perfecting that recipe for oatmeal chocolate chip cookies, she's always got her hands and mind busy."

Emily Rogers
Woodstock, GA

LISA LILLIEN, Voted Diets Be Damned Domestic Goddess Geek

Better known to the weight-loss masses as "Hungry Girl," Lisa Lillien has developed a health food empire—complete with books, recipes, and numerous product tie-ins—based on the idea that losing weight doesn't have to be an uphill battle that drives you bonkers. Instead, you can feed the rumbly in your tumbly *and* drop some lbs at the same time. Lisa has dispensed her nutritious pearls of wisdom on everything from *The View* to the *Rachael Ray* show, and next up is her own *Hungry Girl* show on the Cooking Channel. Plus, Lisa gets major gluten-free brownie points for deciding to build a life with fellow geek—and *iCarly* creator—Dan Schneider. If that doesn't call for some "random dancing," I don't know what does.

KELLY WEARSTLER, Voted Unexpectedly Stylish Domestic Goddess Geek

Fans of *Top Design* probably recognize Kelly Wearstler as one of the show's beloved judges primarily known for her over-the-top getups, nasal vocal intonation, and a mane that's frizzier than a bichon frise—or Twisted Sister's Dee Snider. However, in interior design circles, she's mainly credited with being single-handedly responsible for reviving and modernizing the style of Hollywood Regency.* In addition to penning books like *Modern Glamour* and *Hue*, Wearstler also designs a variety of textiles, including rugs and wallpaper. Part of what makes Kelly so dynamic is that she proves that everyday life can be glamorous—even if you don't have millions in the bank or Bentleys in the garage. By simply incorporating a little glitz into your current decor, you can feel like the queen of your castle.

> "Each room deserves dignity, respect, and a healthy dose of laughter."
> —interior designer Kelly Wearstler

HALL OF FAME: MARTHA STEWART

Yes, ladies and HGTV-obsessed gentlemen. We can't even begin to discuss domestic divas without talking about the one, the only, *the* Martha Stewart. With her refined yet accessible approach to cooking, party planning, and all-around entertaining, Stewart introduced the world to her "good things" in the early 1990s, and the art of creative living hasn't been the same since. The former model first embraced her inner goddess when she started a small catering business out of her kitchen in Westport, Connecticut, and it wasn't long before she was recognized by the culinary elite.

Stewart's first book, *Entertaining*, shot to the top of the *New York Times* best-seller list and quickly spawned an entire series of homemaking titles. The party-planning masses demanded more Martha, so she launched her own magazine, *Martha Stewart Living*, and companion television series. Out of all these ancillary ventures, an empire was born. The marketplace was soon flooded with Martha-endorsed cookbooks, talk

---

* This particular design style, originally made popular in the 1930s by decorators like William Haines and Dorothy Draper, is categorized by oversized furniture, glossy colors, and glamorous hardware. Example of this art deco aesthetic can be seen at the Viceroy Hotel (in Miami), Maison 140 (in Beverly Hills, California), and Bergdorf Goodman's BG restaurant (in New York City), all of which were designed by Wearstler.

shows, publishing ventures, and textile lines for various department stores. These days, it's hard to set a table or fold a fitted sheet without wondering how Martha would do it. It's pretty safe to say that no one's had more of an impact on domestic goddesses than the artist formerly known as federal inmate no. 55170-054—and that's a good thing.

FRENEMIES

Although *Martha Stewart Living*—otherwise known as the bible of how-to—commands an audience of over 11 million readers with each issue, there still exist some haters out there who would rather eat an apple than use said fruit as part of an elaborate place card setting for a summer night's courtyard cocktail party. Crazy, right? Frenemies like this might be hard to spot—unless your daily Facebook feed shows them joining the PWHC Group.[*] But here's a heads-up on which domestic downers might attempt to rain on your perfectly planned—and decorated—parade.

✔ Junk-food junkies who won't eat anything unless it includes trans fat, refined grains, salt, or high-fructose corn syrup.

✔ Hoarders.

✔ A guy who references "scrapbooking" while hooking up.[†]

✔ Know-it-alls who immediately launch into a "but is it art?" discussion after walking through a contemporary art exhibit.

---

[*] People Who Hate Crafts.

[†] "Scrapbooking" usually refers to the act of compiling keepsake albums—or "scrapbooks"—but if a guy mentions "scrapbooking" before, during, or after a trip to Makeout City, he's referring to a *whole* different kind of act—one that doesn't include cardstock and glitter pens. Let's just say it involves a self-produced adhesive, a picture of a hot chick (who isn't you), and a heap of humiliation. In other words, this is *not* something you want to do on your next Crafturday.

- ✔ Eccentrics who wear holiday or Cosby sweaters unironically.
- ✔ Whoever invented lava lamps, shag carpeting, and the high heel shoe chair.
- ✔ Self-proclaimed artists who use paint-by-numbers kits.
- ✔ Anyone whose abode was nominated for—or featured on—the Style Network's *Clean House: Search for the Messiest Home in the Country*.

GEEK LOVE

Much like the Greek goddess Hestia,* our domestically inclined geek sisters are definitely passionate about nesting and all the creative exploits involved in keeping a handsome homestead; however, as opposed to their Olympian counterpart, they are not celibate . . . at least not by choice. After all, love is what makes a house a home, which is why these geek gals are always open to meeting an extra set of hands to help with whatever art/construction/cooking/decorating project they're working on—and if those hands happen to be attached to someone who looks like cutie patootie Scott McGillivray from HGTV's *Income Property*, all the better!

If you're a domestic goddess geekette with an undying interest in assembling, installing, and remodeling, odds are you're probably attracted to a burly, rugged handyman† who can throw you against a wall in a fit of passion and then spackle, sand, and repaint the damaged area. On the other hand, if the only pair of work boots you want to see at the front door are

"I think there's a misconception that a girl being good at domestic stuff somehow makes her submissive and not empowered. Being like, 'I turned this unflattering extra-large Jawbreaker shirt I found on e-Bay into a sweet dress' is definitely not regressive—it's just super-cool."

Magdalena Burnham New York, NY

"She's crafty— and she's just my type." —from the Beastie Boys' "She's Crafty," on *License to Ill*

---

* Goddess of the hearth.
† After all, how else do you explain a tool like Ty Pennington being voted one of *People Magazine*'s "Sexiest Men Alive"? It's surely not because of his puka shell necklaces and frosted tips.

your black-and-gold glitter-flecked Doc Martens, then you're better off with a guy who's a painter, a decorator, or a graphic designer. You know, someone with softer, more delicate hands and a fragile build. However, watch out for supermasculine dudes who fit the alpha male bill. You know the type: competitive, aggressive, and physically superior. That stuff is perfect for a bathroom demo, but not for a romantic relationship.

THE PERFECT MATCH FOR A DOMESTIC GODDESS GEEK GIRL . . .

❏ Looks like Nate Berkus, Carter Oosterhouse, or Jamie Oliver.

❏ Owns an apron and isn't afraid to wear it.

❏ Expresses his love of fonts, especially Helvetica and Georgia.*

❏ Frequents local farmers' markets.

❏ Will read directions—sometimes *twice*, if the item is from Ikea—before attempting to put together a piece of furniture or electronics equipment that says: "Assembly required."

❏ Knows how to sew on a button.

❏ Owns at least three Threadless T-shirts. Bonus points if he's submitted a design idea. Cherry-on-top bonus points if his design was selected and printed as an exclusive Threadless tee.

❏ Offers to help you clean your apartment so he has another excuse to use his Dyson Ball vacuum.

helvetica

* Caveat emptor: beware of any dude who loves Comic Sans. That font screams, "Lives in his mom's basement and runs a Jennifer Love Hewitt fan site."

"CAN'T WE ALL JUST GET ALONG?"

Good help is hard to find—and so are good manners. Sure, the Emily Post Institute still exists to ensure that common courtesy doesn't become extinct, but some new rules of etiquette have cropped up since Post wrote her first manners manual in 1922. Got a question about the dos and don'ts of decorum? Here are the answers every geekette should know to be considered a hostess with the mostess.

"Elbow grease is the best polish."
—English proverb

Q: *How far in advance should you send an invitation or an Evite for an occasion?*
A: It really depends on the occasion. If you were attempting to reserve tickets for the first showing of *The Hobbit* in IMAX, then I'd say at least three to four months ahead. On the other hand, if you're trying to get a head count on how many people will be coming to your weekly *True Blood* viewing party, then I'd say three days ahead—unless you need to plan out how many po' deceased boy sandwiches to make for the affair, in which case five days would suffice.

Q: *I recently received a Facebook friend request from a former high school classmate. We were in the same circle of friends, but I never really liked her that much. I don't want to hurt her feelings, but I also don't really want her to be in my social network. Should I confirm or ignore her?*
A: This is a common netiquette question. Personally, I like to keep questionable requests in my queue for as long as possible. I call this section of my profile "Confirmation Purgatory." Consider it a holding area for acquaintances, coworkers, and other random contacts who don't necessarily *need* to know your personal business or, worse yet, see pictures of you dressed as Lady Gaga from Halloween last year.

Q: *Seriously, who should pay for what on a date?*

A: This is a tricky one. When an invitation is worded "Would you like to try the new Michael Symon restaurant with me on Saturday night?" then the person asking is expected to pay, regardless of gender. If two people both decide to, say, see the latest Marina Abramović exhibition together, then each person would pay his or her own way. However, if both people are delightfully cash-poor, just plan something free and fun so you can avoid the predicament altogether. Worst case? Offer to chip in (and mean it) and see what happens.

> "Most people prolly think of the old-school, '40s/'50s stereotype of women: she cooks, cleans, has babehs, takes care of the babehs, and does womanly things rather than going out and living her dreams of becoming a movie star, a famous singer, etc. [Domestic goddess geeks] do this because this *is* their dream; they aren't doing it to follow the stereotype!"
>
> *Sidni Zimmer*
> *Aurora, CO*

Q: *I have a friend who sends emails in ALL CAPS and insists on asking questions in the subject line—no matter how long the inquiry—and then leaves the body of the message blank. This drives me batty! How can I curb his poor e-communication behavior?*

A: Using all caps in emails is equivalent to shouting. Unless your friend is in need of some major anger management, I'd tell him to lower his tone—and his letter case. The only thing *more* irritating than uppercase-ridden emails is an interminable subject line, which is why I would take said friend off your buddy list STAT. Hell, even Miss Manners knows when to say uncle.

Q: *Whenever I go to a restaurant with my bestie, he/she insists on putting his/her cell phone on the table and proceeds to text throughout the entire meal. This is totally rude, right?*

A: Yes, but maybe it's your bestie's way of telling you that you're a boring dinner companion. Kidding!* You have every right to tell your friend to put the phone away until *after* your meal—unless your BFF is a doctor with a patient waiting on a kidney or something. Then you should (1) grab food with your friend on a night when he or she *isn't* on call or (2) quit your whining, sit quietly, and finish your mesclun salad. Did *you* save anyone's life today?

---

* Unless you *are* a boring dinner companion.

REQUIRED READING

Dwell

Elle Decor

Juxtapoz

Lonny Magazine

Make Magazine

Martha Stewart Living

ReadyMade Magazine

Real Simple

WEB BOOKMARKS

ApartmentTherapy.com
AustinCraftMafia.com
Craftster.org
CraftStylish.com
CraftZine.com
DIYNetwork.com
Etsy.com
GeekCentralStation.Blogspot.com
GeekCrafts.com
GetCrafty.com
Instructables.com
RenegadeHandmade.com
Re-Nest.com
SuperNaturale.com

> "The domestic goddess geek girl is willing to try anything. She could take apart a car engine, assemble the entertainment center (wires and all), take all of the old newspapers in the house and mold them into a giant sculpture of a uterus, knit a Zelda costume for her best friend's baby, and attempt to cook an authentic Indian meal. This girl is the master of her surroundings, and can probably out-skill any man on her block."
>
> *Lindsay Hutton*
> *Blaine, MN*

BAZAAR DESTINATIONS

Bazaar Bizarre (bazaarbizarre.org): Boston, Cleveland, San Francisco

Craftin' Outlaws (craftinoutlaws.com): Columbus, OH

Crafty Bastards (washingtoncitypaper.com/craftybastards): Washington, DC

Detroit Urban Craft Fair (detroiturbancraftfair.com): Detroit, MI

Indie Craft Experience (ice-atlanta.com): Atlanta, GA

Pile of Craft (charmcitycraftmafia.com): Baltimore, MD

Renegade Craft Fair (renegadecraft.com): Austin, Brooklyn, Chicago, Los Angeles, San Francisco

Rock 'N' Roll Craft Show (rocknrollcraftshow.com): St. Louis, MO

Urban Craft Uprising (urbancraftuprising.com): Seattle, WA

MUST-SEE TV

Anthony Bourdain: No Reservations
The Barefoot Contessa
Chopped
Clean House
Design Star
Dinner with the Band
The Fabulous Beekman Boys
Flipping Out
House Hunters
Hungry Girl
Iron Chef America
Property Virgins
Samantha Brown: Great Weekends
Top Chef
Top Design
Work of Art: The Next Great Artist

Seven
MISCELLANEOUS GEEK

Much of being a geek is feeling like you don't *quite* fit in, so it's only natural that this book should include a chapter for geekettes who didn't find a kindred spirit in any of the above caricatures. Plus, much like Pinkberry franchises or the Beckham offspring, new types of geeks are popping up all the time. So consider this the geek kitchen sink chapter—but in a good way.

TECH GEEK GIRL

WHAT SHE'S ALL ABOUT

You don't have to be a full-time computer programmer or software engineer to be a tech geek girl, but having a passion for technology and Internet culture is kind of an inarguable prerequisite. Tech geek girls tend to thrive on intelligence, data, and the constant need to gather information. Not only do they want to *understand* how something works, they also want to know how to *apply* their newfound knowledge to various situations. There's no greater thrill for these open source–loving[*] she-geeks than to get their hands on the latest technological innovation—and master its software—before anyone else. That's why neither snow nor rain nor heat nor gloom of night keeps these women from the swift apprehension of the newest iPad, NOOK, or Xbox 360 Kinect controllerless console.

WHO SHE ADMIRES

Danese Cooper: Dubbed the "Open Source Diva" and the "Wiki Witch of the West," Cooper is probably best known for her programming work at Apple, Intel, Sun Microsystems and, most recently, Wikimedia. She's a staunch open-source advocate and a sought-after speaker who believes the only way to have more female coders is to teach girls to code at an early age. In other words, stop whinging[†] and start doing!

Sara Chipps: Chipps is the cofounder of Girl Develop IT, a series of programming classes—in subjects ranging from HTML/CSS 101 to Introduction to OOP, Databases, and Data-Driven Applications—that are targeted specifically to females. The goal is to narrow the gender gap in software development and teach

---

* Any type of software whose source code is available to the public and open to modification and redistribution without paying a fee to the creator.

† British slang for "to complain obsessively and repetitively." Oddly enough, Little Whinging is also the name of a fictitious, humdrum town in Surrey, England, where Harry Potter lives with his extended family, the Dursleys, before being shipped off to the Hogwarts School of Witchcraft and Wizardry.

girls that anyone can speak JavaScript (no Y chromosome necessary).

Nerd Girls: If you don't think engineering is a feminine profession, think again! Nerd Girls are a group of badass gals who just so happen to be science, technology, engineering, and math students. Started by Dr. Karen Panetta, a professor of electrical and computer engineering at Tufts University, as a way to empower female engineering students, Nerd Girls is all about embracing the left side of the brain and using scientific knowledge to help solve various environmental and technological problems.

HISTORY'S FIRST PROGRAMMER: AUGUSTA "ADA" BYRON KING, COUNTESS OF LOVELACE (December 10, 1815–November 27, 1852)

Lord Byron's daughter Augusta (or Ada, as her friends called her) was a rather sickly girl. She spent most of her childhood on bedrest while battling a variety of ailments ranging from headaches that caused blurred vision to paralysis induced by measles. However, her illness-imposed social exile allowed her to concentrate on her studies and discover her outstanding aptitude for math and science. Through her propensity for mathematics, she met Mary Somerville, a renowned nineteenth-century scientific researcher and author, who then introduced her to Charles Babbage, the "father of the computer."

Babbage and Lovelace soon started corresponding both personally and professionally, and it wasn't long before he asked the math whiz to submit notes on his proposed machine, otherwise known as the Analytic Engine—or "computer." In other words, Lovelace—or, as Babbage dubbed her, "The Enchantress of Numbers"—inadvertently created the world's first computer program, which suggested an algorithm that would calculate Bernoulli numbers.*

HER FAVORITE THINGS

Bands: LCD Soundsystem, Nine Inch Nails, deadmau5

Books: *Predictably Irrational: The Hidden Forces That Shape Our Decisions* by Dan Ariely; *The Best of* Make by editors Mark Frauenfelder and Gareth Branwyn; *Don't Make Me Think: A Common Sense Approach to Web Usability* by Steve Krug

Gadgets: Steampunk USB drive,† Roomba, Intelliscanner‡

Magazines: *Wired, Fast Company, Popular Science, Macworld, Make*

Movies: *WarGames, Hackers, Pirates of Silicon Valley, The Matrix, WALL-E, The Social Network*

Websites: ThinkGeek.com, CNET.com, Gizmodo.com, SlashDot.org, Geek.com, Apple.com, BlogHer.com

---

* In mathematics, Bernoulli numbers are a "sequence of rational numbers with deep connections to number theory." Whatever that means.
† According to *New York Times* writer Ruth La Ferla, "steampunk" can be described as "a subculture that is the aesthetic expression of a time-traveling fantasy world, one that embraces music, film, design, and now fashion, all inspired by the extravagantly inventive age of dirigibles and steam locomotives, brass diving bells and jar-shaped proto-submarines." In simpler terms, it's when modern objects and technologies get a pre-Victorian makeover.
‡ Barcode reader.

WHAT YOUR PDA SAYS ABOUT YOU

What does your PDA say about you and your level of geekocity? Let's take a look at some of the top smartphones on the market and compare them to the personality traits of their potential owners. *"Can you hear me now?"*

BLACKBERRY

You are a serious person with a serious profession who should seriously be taken seriously. In all seriousness, BlackBerry users tend to be high-powered professionals who are more concerned with checking the latest price on their commodity ETF than viewing "Kittens Inspired by Kittens" on YouTube. BlackBerry users are often driven and success-ful and excel at the art of multitasking; however, they also tend to be easily distracted and incapable of holding more than a forty-five-second conversation without taking a hit off their CrackBerry.

DROID

You are quirky, strong-willed, and just *slightly* left of center. You're also extremely loyal, whether it's to friends, family, or a particular mobile service provider. Once you find some-thing you love, you become obsessed instantly. Labels and brand names don't easily impress you, especially because you work hard for the money. (So hard for it, honey.) Af-ter all, you've always felt more comfortable at Target than Barneys, and you don't care who knows it. Your only pet peeve? Convincing people that the Droid isn't just a phone for dudes. Droidettes love it, too. *Geez.*

iPHONE

You are a laid-back intellectual and an independent thinker. You thrive on discovery and always want to be on the ground floor of any trend, movement, or subculture. (For example, you swear that you were rocking braided bangs long before Lauren Conrad ever did. You also recognize the songs in all of the iPod ads without having to Google them.) Whether you're admiring one of Jeff Koons's subversive sculptures or watching the latest episode of *The Real Housewives of New Jersey*, you're okay with straddling the line between being highbrow and lowbrow. So long as you have half a battery, the latest version of Angry Birds, and the Urbanspoon app, all is right with the world.

PALM PIXI

Sorry, toots, but you need to join the twenty-first century and get a *real* smartphone.

WHERE SHE HANGS OUT

The Apple store, coffee shops with free Wi-Fi, various tweet-ups

WHO SHE'S CRUSHING ON

A guy with brains (like Twitter cofounder Biz Stone), bravado (like Vimeo cofounder Jakob Lodwick), or big ideas (like Tumblr cofounder David Karp)

THE TAO OF STEVE JOBS

An apple a day might keep the doctor away, but boasting your Mac love will definitely usher in a wave of eligible male geniuses

—or at least the night manager at the Genius Bar. But why is Apple *the* go-to geek brand of choice? Let's talk about it.

1. The Apple Keynote Speech.

For a tech geek girl, watching the Apple Keynote address is akin to a Browncoat* seeing Nathan Fillion order a tall Americano at the local Starbucks: you start to sweat profusely, butterflies start whirling around in your stomach, and so many adrenaline-induced endorphins are released during the experience that if you were to walk outside and come upon a little kid trapped under a car, you could probably lift the vehicle over your head with one hand, grab the child with the other, and then live to tweet about the whole ordeal. Part PowerPoint, part pep rally, the Apple Keynote is where you find out what piece of brilliant technological equipment is going to make your current piece of brilliant technological equipment obsolete. It's kinda like the tech geek Super Bowl—minus all that pesky athletic activity.

2. Founder Steve Jobs Is Hotter Than Microsoft's Bill Gates.

Okay, so Jobs isn't Johnny Depp or anything, but look at the alternative. Plus, Jobs wears a black turtleneck like nobody's business. Can the same be said about Gates? I think not. (Sorry for the uncomfortable visual. *Shudder.*)

3. Macs Are Prettier Than PCs.

Looks aren't everything, but it's hard to deny how sleek and stylish Macs are compared to their clunky PC counterparts. Hell, even the iPod makes the Zune (R.I.P.) look like an MP3 equivalent of Zach Morris's brick-sized cell phone on *Saved by the Bell*. Tech geek girls like their computers like they like their men: thin, reliable, and virus-free.

---

* Fan of the now-canceled space western TV series *Firefly*.

FASHIONISTA GEEK GIRL

WHAT SHE'S ALL ABOUT

Whether it comes from fitting Barbie into her teeny, tiny open-toed heels or anxiously watching her mother sew together new curtains for the breakfast nook, the fashionista geek girl usually develops a passion for fashion at a pretty early age. She'll spend most of her free time following the trends, but don't get it twisted: this she-geek subsection is fiercely independent and visionary. Clothes become more than just mere objects to cover your body; they express who you *are*—or perhaps who you *wish* you could be. These geek girls are authors, really. But instead of using words to tell a story, they use clothing.

WHO SHE ADMIRES

Tavi Gevinson: Known for her frenetic fashion sense[*]—which blends together high-end couture with whimsical thrift-store finds—and keen eye for up-and-coming trends and designers, Tavi has become the voice of a new generation of fashion lovers through her blog, *The Style Rookie*. Oh, and did I forget to mention that she's barely out of middle school? Wise beyond her years, Tavi has earned both critical praise for her honest and eloquent reporting (*Fast Company* listed her in its "100 Most Creative People in 2010") and some pretty high-profile friends (Marc Jacobs is probably on her phone's friends and family plan). She has joined forces with media mistress

---

[*] You can often find Tavi sitting front row at New York Fashion Week, wearing an oversized bow atop her head. Isabella Blow, the late muse of hat designer Philip Treacy, would be proud.

Jane Pratt to launch a new *Sassy*-esque magazine aimed at like-minded independent (and ultrachic) ladies.

Kelly Cutrone: You might not guess it from her signature stark, all-black wardrobe, but Cutrone is a style maven like no other. She's the founder of People's Revolution—a fashion and lifestyle branding, PR, and marketing firm—and she's responsible for putting together the hottest runway shows all over the world, in addition to turning designers from inaccessible artists to household names (and labels). Not shy to speak her opinion (or rip Whitney Port a new one, like she did so often on *The City*), Cutrone is one of the most respected (and celebrated) women in fashion because she's not afraid to be real and honest, two things that aren't often found in an industry of fantasy and make-believe.

Grace Coddington: Longtime editor in chief Anna Wintour might take up most of the *Vogue* spotlight, but everyone knows that the real star of the fashion show is creative director Grace Coddington. She was originally a model for *Vogue* in her twenties, but her career in front of the camera came to a halt when a car accident left her disfigured. However, instead of abandoning the industry, she decided to focus on the editorial side of the business and pursue a career in styling and production. As seen in the fantastic documentary *The September Issue*, Coddington differs from Wintour in almost every way: Wintour is always dressed to the nines in her Chanel suits and pageboy haircut, while Coddington often looks like she just rolled out of bed, shuffling around the office in worn-out clogs with her hair a fiery mess—and that's why fashionista geek girls worship her! Just because she doesn't look the part of a *Vogue* staffer doesn't mean she has any less love for the biz. If anything, it goes to show you can't judge a magazine editor by her cover.

O-M-G! THAT'S BANANAS!

There is no bigger fashion freak than Rachel Zoe—all eighty pounds of her. Love her or hate her, it's hard to ignore that the girl's got sublime taste for couture and catchphrases. Sure, "Kills it" and "Shut it down" have become the new "That's hot" and "Wazzup," but there are actually a ton of one-liners (aka Zoe-isms) that didn't *quite* catch on. See which ones Bravo left on the cutting-room floor. I like to call these rejects "No-isms."

Discarded catchphrase: "Too much sodium."
Translation: "What's with the salty attitude?"

Discarded catchphrase: "Who farted?"
Translation: "That outfit stinks."

Discarded catchphrase: "Friend request denied."
Translation: "No."

Discarded catchphrase: "You're under arrest."
Translation: "That outfit is so fierce, it's illegal."

Discarded catchphrase: "My left arm just went numb."
Translation: Either (1) "That outfit is so ferosh, I'm totally having a heart attack right now" or (2) "Yes, that outfit is ferosh, but I think I'm actually having a heart attack right now so can someone, like, call 911 or something? Oh, and ask if the ambulance can stop at Starbucks and pick up a trenta nonfat soy latte on the way, mkay?"

HER FAVORITE THINGS

Bands: Lady Gaga, Yeah Yeah Yeahs, Florence + the Machine, M.I.A., Dead Weather, The Sounds

Books: *Style A to Zoe: The Art of Fashion, Beauty & Everything Glamour* by Rachel Zoe and Rose Apodaca; *Avedon Fashion 1944–2000* by Carol Squiers, Vincent Aletti, Philippe Garner, Willis Hartshorn, and Richard Avedon; *The Fairchild Dictionary of Fashion* by Charlotte Mankey Calasibetta and Phyllis Torora; *What to Wear, Where: The How-To Handbook for Any Style Situation* by Hillary Kerr and Katherine Power

Designers: Alexander McQueen, Vivienne Westwood, Rodarte, Chanel, Betsey Johnson, Diane von Furstenberg, Marc Jacobs

Magazines: *Vogue, Nylon, Elle, Teen Vogue, V, Flaunt, Lucky, Women's Wear Daily, i-D, Dazed & Confused*

Models: Alexa Chung, Daisy Lowe, Agyness Deyn, Crystal Renn, Coco Rocha, Lara Stone

Movies: *The Devil Wears Prada, The September Issue, Valentino: The Last Emperor, Clueless, Factory Girl, Funny Face, Coco Avant Chanel*

Websites: Style.com, WhoWhatWear.com, Naag.com, TheStyleRookie.com, WWD.com, Polyvore.com, Weardrobe.com, Fashionista.com

WHERE SHE HANGS OUT

Secondhand stores, sample and trunk sales, in line for the premiere of the newest couture collaboration at H&M, Woodbury Commons Premium Outlets mall (in Central Valley, New York)

WHO SHE'S CRUSHING ON

A guy who's eccentric (like photographer Terry Richardson), a bit mod (like musician-producer Mark Ronson), or deliciously delicate (like actor Nicholas Hoult)

POLITICAL GEEK GIRL

WHAT SHE'S ALL ABOUT

Everyone's favorite English philosopher named after a pork product, Sir Francis Bacon, coined the phrase "Knowledge is power," and these three small words perfectly sum up the mantra of the political geek girl. Not only are these ladies driven by the quest for knowledge, they also aim to use the constant influx of information to help make a difference in the world. They strive to be advocates and activists; thus they often possess a pretty rigid set of values and ethics. (Some might call them stubborn or obstinate. Not *me*, of course, but some.) In an ideal world, everyone would see things *their* way. Unfortunately, their vision for a Utopian society doesn't always mesh with the views of others, which is why these passionate protesters always stand out in a crowd—mostly because they're the ones spouting political buzzwords like "Obamacare," "earmarks," and "Michael Moore" at obscenely loud volumes.

WHO SHE ADMIRES

Rachel Maddow: If political geek girls aren't glued to MSNBC weeknights at 9:00 p.m., then they've definitely got a season pass set up on their TiVo. No matter what they have to do, missing *The Rachel Maddow Show* isn't an option. Openly gay and undoubtedly liberal, Maddow approaches the day's top political headlines with refreshing candor, insight, and the kind of deadpan delivery taught at the Jon Stewart school of news anchoring. Her outspoken opinions on hot topics like Fox News and the Tea Party have made her a somewhat polarizing figure, but it's hard not to respect her spark and sass—especially when she's calling Bill O'Reilly a "racist-baiting f***."

Gloria Steinem: No matter how you feel about her political agenda, Steinem is inarguably a feminist icon, and the entire geek girl scene would look very different without her groundbreaking efforts. In the 1960s, she was on the front lines of the women's liberation movement and sexual revolution. More than anything else, Steinem has always spoken in favor of humanism and the desire to live in a society where *everyone* was treated fairly, regardless of color, gender, creed, etc. Over the past four decades, she's founded more nonprofits than I can mention in this small space (Women's Media Center and Women's Action Alliance, just to name a few), in addition to *Ms.* magazine, which remains an active watchdog of feminist culture.

Meghan McCain: This political blogger might be Republican royalty,* but she's not afraid to express her liberalism on social issues like abortion (she doesn't believe in abstinence-only sex education), gay marriage (she's all for it), and immigration (she spoke out against the Arizona act, which her father supported). Meghan first embraced the public platform when she started blogging during her father's 2008 presidential campaign; two years later, she

---

* She's the daughter of U.S. Senator John McCain.

released a book about her time on the trail called *Dirty Sexy Politics*. She's currently a contributor to *The Daily Beast*, where she writes about everything from country crooner Miranda Lambert to the union protests in Wisconsin. Finally, a Republican role model who doesn't spout nonsense like, "I'm a Christian first, and a mean-spirited, bigoted conservative second, and don't you ever forget it."*

SUFFRAGETTE CITY: WHAM, BAM, THANK YOU, MA'AM!

If you're like me, one of your favorite songs from *Mary Poppins* is "Sister Suffragette." However, if you're like me, you were too blinded by Glynis Johns's straw boater hat and fancy sash to understand what the tune was *really* about. In fact, if not for the activism of the suffrage movement, we geek girls wouldn't be able to do things like own property, run for office, or pay taxes. Okay, we could've gone without that last one, but these civil rights sisters deserve to be celebrated. As Mrs. Banks would say, "Well done, sister suffragette!"

1776 – Before founding father John Adams drafts the Declaration of Independence, his wife, Abigail, asks him to "remember the ladies." He agrees and then specifies in the doctrine "all men are created equal." This won't be the first time a husband exhibits short-term memory loss.

1787 – Women lose the right to vote in all states except New Jersey, which will eventually disenfranchise women ten years later.

---

* Can someone *please* put a muzzle on Ann Coulter? I mean, really.

1792 – Philosopher Mary Wollstonecraft publishes *A Vindication of the Rights of Women* in England. The U.S. population prepares for the fact that everything cool will happen in Britain first.

1833 – Ohio's Oberlin College becomes the first coeducational university in the United States. Notable future female alumni will include *Sassy* editor in chief Jane Pratt, singer Liz Phair, and Yeah Yeah Yeahs frontwoman Karen O.

1848 – The first Women's Rights Convention is held in Seneca Falls, New York. It was probably like the Michigan Womyn's Music Festivals, minus the Ani DiFranco cover bands and seminars on lesbian polyamory.

1859 – Rubber condoms are introduced as a means of contraception. These early prototypes will provide future inspiration for the infamous sex scene between Leslie Nielsen and Priscilla Presley in *The Naked Gun*.

1867 – The Fourteenth Amendment passes Congress, defining citizens as "male." Civil rights leader Susan B. Anthony reacts by forming the Equal Rights Association, which fights for universal suffrage.

1870 – The Fifteenth Amendment is ratified; it states, "The right of citizens of the United States to vote shall not be denied or abridged by the United States or by any State on account of race, color, or previous condition of servitude." However, when women take to the polls, their votes aren't counted.

1878 – A women's suffrage amendment is introduced to Congress. It will take almost thirty years to pass.

1920 – The Nineteenth Amendment is ratified by Congress. It promises that no U.S. citizen will be denied the right to vote on account of sex.

HER FAVORITE THINGS

Bands: Against Me!, Bob Dylan, the Clash, Arcade Fire, the Mars Volta, Bruce Springsteen, Sleater-Kinney, Le Tigre, the Gossip

Books: *A People's History of the United States* by Howard Zinn; *The Daily Show with Jon Stewart Presents America (The Book) Teacher's Edition: A Citizen's Guide to Democracy Inaction* by Jon Stewart and the writers of *The Daily Show; Our Dumb Century* by Scott Dikkers and the editors of *The Onion; 48 Laws of Power* by Robert Greene

Movies: *Thank You for Smoking, Wag the Dog, An Inconvenient Truth, Bowling for Columbine, All the President's Men, Food Inc., Inside Job*

Magazines: *Mother Jones, The New Yorker, The Nation, The Weekly Standard, Slate*

Websites: Salon.com, DrudgeReport.com, TheDailyBeast.com, TheHuffingtonPost.com, Wonkette.com, FactCheck.org, Truthdig.com

WHERE SHE HANGS OUT
Dim sum restaurants, city hall, basement and/or living room punk-rock shows

WHO SHE'S CRUSHING ON

A guy who is steadfast in his beliefs (like Stephen Colbert, host of *The Colbert Report*), can wear a bow tie like nobody's business (like Fox News political correspondent Tucker Carlson), or doesn't take himself *too* seriously (like *The Daily Show* host Jon Stewart).

RETRO GEEK GIRL

WHAT SHE'S ALL ABOUT

You know that saying, "Those who don't learn from history are doomed to repeat it"? Well, that'd be totally okay with our retro geek girl. In fact, if given a choice, she'd probably prefer to take a trip back to simpler—and more stylish—times when crocheted ponchos weren't mocked and facial hair wasn't ironic. Generally speaking, these gals tend to be gregarious and self-confident and have an unflinching zest for life, much like Janeane Garofalo's Vickie character in *Reality Bites*, who would rather bask in nostalgia than enjoy anything in the present tense. Many even embrace the attitude, "If it's not fun, why do it?" (I'm still not sure if that's because these girls just wanna have fun or because those Ben & Jerry's bumper stickers are awful catchy.) But in a world where many people romanticize the future, this group of she-geeks is still smitten with the past.

WHO SHE ADMIRES

Chloë Sevigny: A big part of being a retro geek girl is acting cool. Just like the tagline for that old-school brand of deodorant Dry Idea, you never wanna let them see you sweat. That said, if cool were a commodity, Chloë Sevigny would be Warren Buffett—or at least Jimmy Buffett. (Goodness knows those Parrotheads have allowed him to live pretty damn comfortably.) Chloë's carefree attitude applies to both her acting choices and her personal style. Whether she's giving fellatio to Vincent Gallo in *The Brown Bunny* or wearing an oversized teal doily to the annual Met Costume Institute Gala, she doesn't care what other people think, and that's what being a retro geek girl is all about.

Ana Calderon: It's hard to flip through any of the galleries on the Cobra Snake without seeing Ana Calderon somewhere in the frame. With her signature black-brimmed hat and red-lipstick smirk, this Kansas-born trendsetter stands out in a crowd. Whether she's DJing or promoting an event in the hipster hood, the party doesn't start until Calderon walks in. She might be queen of the L.A. scene now, opening for bands like Warpaint and Interpol, but this retro trend-spotter originally got her start by doing marketing at record labels like Hopeless Records and then Dim Mak, owned by party boy (and eventual mentor) Steve Aoki. Bowler hats off to you, girl!

The Like: The founding members of the Like—vocalist-guitarist Elizabeth "Z" Berg and drummer Tennessee Thomas—grew up in the 1980s as the daughters of music-biz royalty,[*] but the girls were inspired by the style and sound of the 1960s more than anything else. They formed the band in their midteens, and soon found themselves playing alongside other local retro-tinged acts like Phantom Planet, Rooney, and Kara's Flowers.[†] The Like's music has yet to resonate with mainstream audiences, but their mod style certainly has. The girls starred in a short film promot-

---

* Berg's father, Tony, worked for Geffen and in the mid-1990s signed artists like Beck and At the Drive-in, while Thomas's father, Pete, is the longtime drummer for Elvis Costello.
† Better known now as Maroon 5.

ing Zac Posen's line for Target, and their influence can be seen on anyone wearing minidresses or hot pants at their shows.

HER FAVORITE THINGS

Accessories: One single feather earring, Ray-Ban sunglasses, turquoise jewelry, batik scarf, bowler hat, ShakeItPhoto iPhone application, ceiling fan pull chain doubling as a bracelet

Bands: Crystal Castles, MGMT, Sharon Jones and the Dap-Kings, Rooney, Hot Chip, Passion Pit, the Strokes, Bat for Lashes, Edward Sharpe and the Magnetic Zeros, Warpaint, Odd Future Wolf Gang Kill Them All

Books: *Terryworld* by Dian Hanson and Terry Richardson; *The Vice Guide to Sex, Drugs and Rock and Roll* by Suroosh Alvi, Gavin McInnes, and Shane Smith; *Dear Diary* by Lesley Arfin; *Party Monster: A Fabulous but True Tale of Murder in Clubland* by James St. James

Movies: *Dazed and Confused, The Virgin Suicides, Valley of the Dolls, Kids, The Big Lebowski, Heathers, Coffee and Cigarettes, Fear and Loathing in Las Vegas, A Clockwork Orange*

Websites: ViceLand.com, TheCobraSnake.com, ImBoyCrazy .com, TheFader.com, ItsCoryKennedy.Wordpress.com, BuiltBy Wendy.com, LastNightsParty.com

20/20 VISION

Dorothy Parker may've said "Men rarely make passes at girls who wear glasses," but that hardly holds true today. In fact, four eyes are better than two when it comes to geek chic. Whether they're prescription or not,* here's a little insight about what your specs say about you.

Browline: "I follow current events and occasionally read the newspaper. By 'read the newspaper,' I mean I loaded the *New York Times* app on my iPad and sometimes I accidentally click on Politics instead of Entertainment. That still counts, right? Oh, and before you say it, I was wearing these frames *waaay* before Harry Crane on *Mad Men*. Ask anyone."

---

* No judgment.

Monocle: "I'm a descendent of Mr. Peanut."

Horn-rimmed: "I can be shy one minute and gregarious the next. I'm just like a box of chocolates: you never know what you're gonna get. Oddly enough, people often mistake me for Forrest Gump. Go figure."

Cat's Eye: "I'm brassy and quirky and I have a soft spot for 1950s pinups. If guys happen to mistake me for a hot librarian, so be it."

Rimless: "Because I'm an artistic soul, I often frequent street fairs, craft fairs, and swap meets. If one girl's trash is another girl's treasure, then I'm the richest gal alive. So what if my parents keep threatening to call *Hoarders*? How much living space does one person really need?" *(Stubs toe on corner of broken Singer portable suitcase record player.)*

WHERE SHE HANGS OUT

Roller derby competitions, themed dive bars and twenty-four-hour diners, vintage stores, the VIP tent at Coachella

AT A GLANCE: WILLIAMSBURG VS. SILVER LAKE

Located more than 2,500 miles apart, Silver Lake (east of Los Angeles) and Williamsburg (in Brooklyn, New York) are both hotbeds of hipster activity. Whether you find yourself living with seven other people in a loft above Bedford Ave. or shacking up with your bike messenger boyfriend in a ramshackle guesthouse off the main drag of Sunset Junction, each neighborhood offers up its own unique attitude.

Not sure which hood is which? Let's break down some of the style elements that separate the 'Burg from the Lake.

MUST-HAVE	SILVER LAKE	WILLIAMSBURG
Outerwear	Vintage jean jacket	Pea coat
Hat	Knit beanie	Fedora
Boots	Minnetonka	Frye
Sunglasses	Nina Ricci Jackie O.	Ray-Ban aviators
Sandals	Thong	Gladiator
Bag	Satchel	Messenger
Dress	Maxi	Mini
Heels	Wedges	Stilettos
Hair	Blond, long, and layered	Brunette, short, and severe
Resale shop	Wasteland	Beacon's Closet

WHO SHE'S CRUSHING ON

A guy who's effortlessly cool (like Julian Casablancas, singer of the Strokes), bat-sh*t crazy (like the Cobra Snake photographer Mark Hunter), or creatively innovative (like Girl Talk musician and DJ Gregg Gillis).

ATHLETIC GEEK GIRL

WHAT SHE'S ALL ABOUT

Typically, geeks might not be the most athletic human specimens, but that doesn't mean they don't like to cheer on their favorite sports teams—or participate in athletics that don't require a whole lot of physical contact. In fact, sports geek girls tend to be pretty competitive. Not only will they yell at the TV if their team isn't winning, but don't be surprised if they talk major smack during your supposedly innocent slow-pitch softball game. However, what these geeks lack in self-restraint, they make up for in leadership, drive, and the motivation of others. After all, winning isn't everything—it's the only thing! Just kidding . . . *sorta.*

WHO SHE ADMIRES

Erin Andrews: Mainstream fans might know Erin Andrews from her impressive stint on *Dancing with the Stars*, where she finished third, but most sports nuts are familiar with the statuesque blonde from her sideline interviews with up-and-coming college basketball, baseball, and football stars. She might be a bombshell, but don't be fooled by her looks—this girl can talk sports with the best of 'em and cites sportscasting greats like Lesley Visser, Michele Tafoya, and Pam Oliver as her inspirations. Erin currently hosts ESPN's *College Gameday* on Saturday mornings and is also a frequent contributor to *Good Morning America.*

Serena and Venus Williams: Even if you don't speak Wimbledon, you probably know the miraculous Williams sisters, two of the most celebrated—and controversial— tennis players ever to step onto the court. Their father, Richard, moved the entire family from Compton to West Palm Beach when Venus and Serena were barely tweens so the girls could hone their tennis skills with famed coach Rick Macci, but Richard eventually took over as their trainer and the prodigies went on to win every tennis title in existence. U.S. Open? *Check.* Wimbledon? *Of course.*

Olympics? *Duh.* Over the years, Venus and Serena have not only added spark and flash to a traditionally stuffy sport; they've also inspired countless young girls to take life by the balls[*] because anything is possible.

Hannah Storm: Seeing that Storm's father was commissioner of the American Basketball Association, general manager of numerous b-ball franchises, and president of the Atlanta Hawks, it's easy to see that sports are in the girl's blood—although it wasn't until she worked in radio after college that she discovered her beat . . . and bitchin' last name.[†] Storm quickly rose through the sportscaster ranks and became the first female host of *CNN Sports Tonight* in 1989. Since then, she's (wo)manned the desk for NBC Sports, *The Early Show,* and ESPN's *SportsCenter,* which she currently co-anchors with fellow sports journalist Josh Elliott. Go girl![‡]

HER FAVORITE THINGS

Bands: Jay-Z, Weezer, Kings of Leon, Incubus, Eminem, classic rock

Books: *The Time Traveler's Wife* by Audrey Niffenegger; *Pride and Prejudice* by Jane Austen; *The Girl with the Dragon Tattoo* by Stieg Larsson; *I Hope They Serve Beer in Hell* by Tucker Max

TV: *Grey's Anatomy, Scrubs, Friday Night Lights, Weeds, SportsCenter, The League, Arli$$, Sports Night*

---

[*] Figuratively speaking, of course.

[†] Born Hannah Storem, she took on the surname Storm while DJing at a rock radio station in Texas.

[‡] Storm penned a book with this *exact* title, which is a guide for parents who want to raise their daughters to participate in sports.

Movies: *Caddyshack, Happy Gilmore, Step Brothers, The Mighty Ducks, Top Gun, Rudy, Remember the Titans*

Websites: ESPN.com, LiveScore.com, DeadSpin.com, SportsPickle.com, BleacherReport.com, TheHuddle.com, RotoWorld.com

WHERE SHE HANGS OUT

Stadiums, sports bars, local gymnasiums that host community dodgeball, touch football, or kickball leagues

WHO SHE'S CRUSHING ON

A guy who's soft-spoken (like Cleveland Indians center-fielder Grady Sizemore) or has All-American looks (like New England Patriots quarterback Tom Brady) or rock-hard abs (like Real Madrid forward Cristiano Ronaldo).

Of course, this is just scratching the surface of the tons of geekettes out there. I'm trying my best to keep track of 'em but I need your help! Hoof it to my website, www.leslie-simon.com, and tell me about the burgeoning geek obsession that's made you a smitten kitten.

GEEK GIRLS UNITE

While our time together might be coming to an end, the real journey is just beginning. In each chapter, you learned about various geek goddesses who redefined gender roles, broke barriers, and entered uncharted territory in their particular areas of expertise. Now it's time for you to take the geek girl baton from them and run—or at least walk briskly—with it.

Acknowledging and accepting your geekiness isn't always an easy task—at any age—especially when you've tried your hardest to fit in and all you seem to do is stand out. When I think back to what I was like in high school, it's hard for me not to cringe and facepalm. I despised most of my friends, boys often hid from me, and no matter how hard I looked, I just couldn't seem to find my niche. One Saturday night toward the end of senior year, my longtime bestie Jessica was having people over because her parents were out of town. The idea of drinking wine coolers in the comfort of her traditional Tudor was much more appealing than congregating outside at other underage drinking spots like the Ravine and the Rock. I drove over by myself because I couldn't find anyone to carpool with, and the moment I walked in the door, I knew the night was going to suck.

I didn't drink or smoke in high school, so people treated me like a nark. In other words, they tolerated my presence but secretly wished I would leave. This one kid we hung out with, Rob, wasn't so stealth about his contempt for me. However, on this particular evening, he was being oddly nice and charming. I should've known this was a setup because Rob was rotten. He was the kid who, in seventh grade, asked all the girls in English class if their hymen was hanging and would ruin every photo op by "hanging brains."* Like I said, rotten. After a couple of hours of chatting me up, he asked when I was leaving and if he could have a ride home. Me being the trusting, gullible sack I was, I said "Sure" and told him my ETD was fifteen minutes. I made the rounds, saying my good-byes, and told Rob it was time to hoof it. It was then that he said he actually didn't need a ride anymore.

I thought nothing of it and walked out alone to my parents' Subaru, which was parked on the street in front of Jessica's house. As I slowly approached the station wagon, I started to notice these white polka dots all over the car. I wasn't more than five feet away when it clicked: I had been egged—and by that little weasel, Rob. The shock wore off the second I heard giggle fits erupting behind me, and that's when I started to feel the punch in my gut and the lump in my throat. First came the tears, then came the rage. By this time, a small crowd of people had gathered on the lawn to see what the commotion was all about. Jessica walked over to try to calm me down, but I was

---

* Dangling your testicles outside of your pants.

crazy mad, like Christian Bale on the set of *Terminator*, so I grabbed her by the shoulders and shoved her aside.

I stormed into the house to find Rob sitting in the den, hysterically laughing. I'm not sure what I said—er, screamed—at him, but whatever it was, it just made him cackle harder. Realizing I had to be home in forty-five minutes and my parents' car looked like an omelet, I raced over to the nearest car wash and prayed that BP's Triple Foam System would eliminate all remnants of the eggs and this entire night. As I drove, I sobbed. My nose ran and my body heaved. Not only had I been humiliated in front of all my so-called friends, but the poor station wagon was one big, egged eyesore. I felt like all the neighboring drivers were pointing and mocking my pain, and that made me cry even harder.

Then, out of nowhere, I heard a voice. It was sweet, soothing, and reassuring. I took a couple of deep breaths and the tears soon stopped. No, it wasn't God. As ashamed as I am to admit, it was Dave Matthews and he was singing "I'll Back You Up." In that moment, his lyrics were all the encouragement I needed to know that everything was going to be okay: "Do what you will, always . . . Walk where you like, your steps . . . Do as you please, I'll back you up." I finally discovered something that would never let me down or abandon me. Something that would never egg my car—or agree to go to homecoming with me only to call two days later and back out because he thought he'd have more fun going with someone else. Something that made me happier than I thought possible. I discovered music, geeked out, and knew my life would never be the same.

More than ten years later, I'm over those high school slights, but I'm still completely head-over-heels in love with music. Sure, my tastes have changed over the years, but I still get butterflies when I hear an amazing song, whether it be for the first or the hundredth time. That's why I'll always be proud to call myself a music geek girl—and I'll never "Say Goodbye" to the boxes of Maxell XLII tapes filled with all the DMB bootleg live shows I collected during my late teens.

WITHOUT A PADDLE

Just because the National Pan-Hellenic Council doesn't recognize the Geek Girls Guild (ΓΓΓ) doesn't mean we can't celebrate the notion of sisterhood. However, depending on where you live, that might be easier said than done. When you reside in metropolitan cities like New York, Los Angeles, or Chicago, the world is your geeky oyster. But what if you live in Stars Hollow, Massachusetts, and you're the only girl at your school who watches *Strangers with Candy* and collects Keith Haring T-shirts? I'm betting you've probably been ostracized by your Abercrombie & Fitch–wearing, Dane Cook–loving peers, right? Or what if you work in corporate America and you're not allowed to prominently display your beloved Dwight Schrute bobblehead? If that's the case, I'm guessing your cubemate probably *isn't* the best audience for your out-of-nowhere quotes about red staplers and TPS reports.

Sure, you might feel like the odd one out, but you are not alone. In fact, most of your sisters are just a couple of steps—or clicks—away. Allow me to offer some pearls of wisdom on how to connect with kindred spirits and make sure the highlight of your weekend *isn't* watching *Medium* with your parents or sitting alone in your apartment while looking up videos of people popping zits on YouTube.* You've probably heard or thought of a lot of these tips before, but if you haven't taken action on 'em, I hope this list inspires you. I know that working on this book has inspired me to read more Neil Gaiman, look into enrolling

---

* Obviously, not my proudest moment.

in classes at Upright Citizens Brigade, and do whatever it takes to make Olivia Munn my new bestie. Sky's the limit, geekettes!

Be Proud of Who You Are.

Haters and frenemies might try to throw shade your way, but don't let it get to you. You've gotta dust that dirt off your shoulders. After all, your geekiness is something that should be respected, revered, and celebrated. Some say, "What doesn't kill us makes us stronger." I say, "What doesn't kill us makes us more motivated to succeed, love life, and then rub our exciting accomplishments in the faces of all the loser bullies we'll inevitably run into at our ten-year high school reunions." I think even Nietzsche would be onboard with that outlook.

Get Active.

No matter what kind of geek you are, get involved and stake your claim in the local scene. That might mean organizing a craft swap for the girls in your dorm or volunteering to be a ticket wrangler at the nearest film festival. Get out there and spread the geek gospel. Become a force to be reckoned with, and earn a reputation for being an astute badass.

Find a Mentor.

I highly recommend finding a "womentor," which is my fancy-pants word for an awesome lady with rad skills who happens to be forging ground in your area of desired expertise. Wanna be a music journalist? Reach out to the female music editor at your local entertainment paper and see if she'll take a look at some of your blog's album reviews. Hoping to run an art gallery someday? Lend a hand at your city or town's next gallery hop and see if you can convince an already established curator to let you walk in her stilettos. Most movers and shakers *want* to help mold and foster the next generation, so all you have to do is ask.

Speak Your Mind and Ask Questions.

Don't be afraid of the sound of your own voice, whether it's on-line or in person. Take the extra time to wait by the merch booth to chat with your favorite band after a concert, or submit a new use for an old thing to *Real Simple* magazine. Not only will this sharpen your social skills, but you never know where you might meet a like-minded sister-friend. She could be in a Stieg Larsson fan forum, or in line for tickets to the Tim and Eric Awesome Tour. Curiosity is at the core of every geek girl, and if you never ask, the answer's always no. Plus, I'm going to let you in on a little secret: everyone *loves* to talk about themselves—especially when it's about their passions, accomplishments, and successes.

Be Patient.

World domination isn't something that happens overnight, un-less your name is Justin Bieber.

Start Your Own Geek Girl Guild (ΓΓΓ) Chapter.

When I first started the ΓΓΓ, my goal was to learn more about our individual passions, unite us as a geektastic girl community, and ultimately celebrate our fantastic uniqueness. Judging from the head count of our first pledge class—108 sorority sisters!—I'd have to say mission *almost* accomplished. I encourage all of you to start your own ΓΓΓ chapter. Recruit your bestie, the gaggle of gals you eat lunch with, or that girl in your philosophy class who had a really interesting take on existence preceding essence.

Meet as often as you want, and try to hold your meetings in a location that really gets the creative juices flowing. (That might be the co-op coffee house or the basement at your parents' house. As long as there's good lighting, proper ventilation, and delicious snacks, you're golden.) The agenda is completely up to you, too! Make it as focused or as dynamic as you want. One week, you might talk about whether you're offended by designer Jonathan Adler selling a pillow with the pro-ana slogan NOTHING TASTES AS GOOD AS SKINNY FEELS needlepointed on it. The next week, you might discuss the cinematic merits—if any—of director Harmony Korine. The possibilities are endless.

Oh, and be sure to let me know when you start a ΓΓΓ chapter because I'd love to keep track of all the burgeoning ΓΓΓ colonies out there.

Appreciate the Geek Girls Around You.

And I'll be the first to start: big ups to all the fangirls, fashionistas, bookworms, indie chicks, craft mavens, and other female misfits out there! Your wild style never ceases to amaze me, and I look forward to seeing how your fabulosity will shape and influence future generations.

Together, we geek girls shall inherit the earth!

ACKNOWLEDGMENTS

I'd like to extend a heaping helping of thanks to my geek guardian angels—my agent Lisa Grubka and my editor Stephanie Meyers. Without their unwavering support and encouragement, I don't think I would've been able to write this book without losing my marbles. Nan Lawson, my beloved illustrator, was such a gem throughout this process and I consider her my own geeky Athena.* I'll also be forever indebted to all the incredible gals who participated in the first rush class of the Geek Girl Guild (ΓΓΓ). Your input has made this book a million times more kick-ass. Sayin'.

I want to send loads of love to my family for putting up with my stress, anxiety, and all-around crazy antics for the past year—hell, for the past thirty-two years. Not even the likes of Amy Sherman-Palladino could script me a better fam. Bear

---

* Goddess of the arts.

hugs and sloppy kisses to Eileen and Jeff Simon, Marilyn Curtiss, Jennifer Curtiss, and the California Gresham-Curtiss clan.

I'm also one extremely lucky geek girl to be surrounded by the most talented and inspiring group of fabulous lady friends. I consider you all my sisters from another mister, and I bow to your awesomeness. Thank you for not turning your backs on me when I postponed brunch, canceled dinner, bailed out on drinks, forgot to call you back, or took a million years to reply to your emails. I love you all so much: Bonnie Dillard, Bridget Gibbons, Gurj Bassi, Jessica Weeks, Kate Cafaro, Lesley Federman, Linsey Molloy, Rachel Lux, Robin Benway, Sara Newens, Debbie Wunder, Christina Johns, Teeter Sperber, and Sarah Saturday.

ABOUT THE AUTHOR

LESLIE SIMON thinks these author blurbs are awkward, trite, and oftentimes a total snooze, so here she is in a nutshell: Leslie lives in NYC, isn't a fan of hot weather, and loves her parents, *Gilmore Girls,* and French bulldog puppies. She's the author of *Wish You Were Here: An Essential Guide to Your Favorite Music Scenes* and coauthor of *Everybody Hurts: An Essential Guide to Emo Culture.* She is currently an editor at MTV.com, where she writes about Justin Bieber more than she thought was humanly possible. If you want to find her in the Twitterverse, follow @redpatterndress and @geekgirlsunite.

BOOKS BY LESLIE SIMON

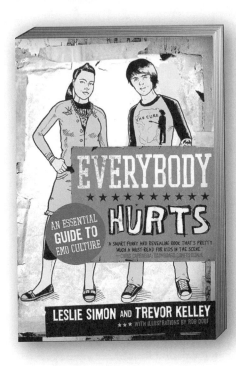

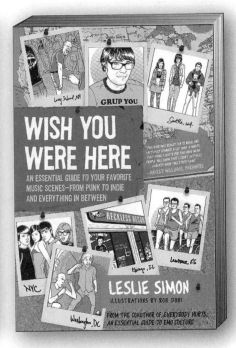

EVERYBODY HURTS
An Essential Guide to Emo Culture

Learn everything you wanted to know about what it means to be "emo," including the culture's music, ideology, and fashion. Filled with original artwork, playlists, movies picks, and more—this book is essential for emo experts and newcomers, and will be a must-read for years to come.

"A smart, funny, and revealing book that's pretty much a must-read for kids in the scene."

—Chris Carrabba, lead singer and guitarist of Dashboard Confessional

WISH YOU WERE HERE
An Essential Guide to Your Favorite Music Scenes—from Punk to Indie and Everything in Between

In this travel guide meets tongue-in-cheek history, take a trip to all the cities—and scenes—that help make up the state of indie rock as we know it.

"Wait a minute. Does giving a quote for this book mean people will know that I didn't actually already know this stuff? Lame."

—Hayley Williams, lead singer of Paramore

itbooks
AN IMPRINT OF HARPERCOLLINS PUBLISHERS

Available wherever books are sold.